101

D1418750

Earth Building

WEEK LOAN

Earth Building

Methods and materials, repair and conservation

Laurence Keefe

Taylor & Francis
Taylor & Francis Group

LONDON AND NEW YORK

First published 2005 by Taylor & Francis
2 Park Square, Milton Park, Abingdon, Oxon OX14 4RN

Simultaneously published in the USA and Canada
by Taylor & Francis
270 Madison Ave, New York NY 10016

Taylor & Francis is an imprint of the Taylor & Francis Group

© 2005 Laurence Keefe

Typeset in Univers by Florence Production Ltd, Stoodleigh, Devon
Printed and bound in Great Britain by TJ International Ltd, Padstow, Cornwall.

British Library Cataloguing in Publication Data
A catalogue record for this book is available from the British Library

Library of Congress Cataloging in Publication Data
A catalog record for this book has been requested

ISBN 0–415–32322–3

Contents

List of illustrations   vii

Acknowledgements   xi

Preface   xii

Part I   Earth construction: methods and materials   1

Introduction: earth as a 'green' building material   1

1 Earth building in Britain – international context and historical background   7

By R. Nother, with an introduction by L. Keefe

British earth building in an international context: a brief history   7

Earth buildings in Britain – types and distribution   10

The distribution of earth buildings in Britain   16

2 The raw material: soils for use in building   30

Earth as a construction material   30

Soil classification in detail: identification, analysis and testing   34

Clays and aggregates, including chalk   43

3 Earth construction 1: preparation and mixing   51

Constructing mass earth and earth block walls   51

4 Earth construction 2: the building process   71

By L. Keefe and R. Nother

Cob wall construction   71

Building with rammed earth   85

Building with earth blocks and clay lump   94

Earth-based finishes for earth walls   96

5 Standards and regulations   101

By R. Nother, with a note on Standards and Codes of Practice by L. Keefe

Introduction   101

Planning legislation, policies and guidance   102

Contents

Building Regulations 103

The Building Regulation Approved Documents 104

Approved Document to support Regulation 7 –
materials and workmanship 105

Approved Document A – structure 107

Approved Document B – fire safety 109

Approved Document C – site preparation and resistance
to moisture 111

Approved Document E – resistance to the passage
of sound 113

Approved Document L1 – conservation of fuel and power
in dwellings 114

Approved Document L2 – conservation of fuel and power
in buildings other than dwellings 114

Technical standards and codes of practice for earth-wall
construction 121

**Part II   The conservation, repair and maintenance of
earth buildings** **125**

Introduction 125

Why preserve earth buildings: the need for conservation? 127

**6 The principal causes of failure in earth walls and how to
recognise them** 131

Inherent defects associated with original construction
methods and materials 131

Problems arising from later alterations and interventions 136

The use of inappropriate materials and ineffective
repair methods 139

Physical damage from external sources 146

Dampness in earth walls and associated building materials 151

7 The diagnostic survey and repair of earth buildings 158

A suggested diagnostic survey procedure for earth-walled
buildings 158

The repair of earth buildings 164

Organisations 188

Bibliography 190

Index 193

List of illustrations

1.1   Typical constructional details of early cob and mudwall buildings   9
1.2   Reconstruction of a 17th century New Forest squatter's cottage   13
1.3   Details of mud and stud construction   15
1.4   Cob and clay-lump walls compared   15
1.5   Regional distribution of surviving earth buildings in Britain   17
1.6   Reconstruction of an 18th century Welsh 'clom' cottage,
      Cardiff   19
1.7   Clay Dabbins. Farmhouse and restored barn, Cumbria   19
1.8   Mud and stud houses at Thimbleby. Lincolnshire   21
1.9   Clay lump building at Bressingham, Norfolk, c. 1850   23
1.10 Render removed from the walls of clay lump cottages in
      Norfolk   24
1.11 Witchert houses and boundary walls at Haddenham,
      Buckinghamshire   25
1.12 Chalk cob house at Amesbury, Wiltshire, 1920   25
1.13 Late Georgian cob town houses in Dawlish, Devon, c. 1820   26
1.14 Early nineteenth century chalk cob farmhouse, Dorset   27
1.15 Early nineteenth century non-conformist chapel, Devon   28
1.16 Five storey rammed chalk houses in Winchester, Hampshire   28
2.1   Clays and aggregates – relative particle size diameter   33
2.2   Particle size distribution (psd) chart   33
2.3   Volume change, consistency and moisture content in soils   38
2.4   Machine for testing the compressive strength of soils   41
2.5   Relationship of compressive strength to moisture content   41
2.6   Typical unconfined compressive test specimens   42
2.7   Partly collapsed cob wall, showing the effects of long term
      weathering   45
2.8   Micro-particles: SEM photographs of clay and chalk particles   46
2.9   Diagram showing electro-static bonding between clay particles   47
2.10 The effects of moisture on the behaviour of clay particles   47
3.1   Suggested optimum particle size distribution curves   54
3.2   Controlling shrinkage by the addition of organic fibres   58
3.3   Traditional method of mixing and placing cob   59

List of illustrations

3.4	Mixing cob using a mechanical digger	61
3.5	Cob drying out on timber pallets	61
3.6	Traditional method of making mud bricks (adobes)	63
3.7	Cob blocks fabricated using a four-gang timber mould	65
3.8	CINVA ram type compressed soil block press	68
3,9	Operating a manual soil block press	69
4.1	Placing cob using a dung fork and lead dressing tool	72
4.2	Method of paring down a cob wall, also showing formwork	73
4.3	Treading and beating cob	74
4.4	Curved cob walls under construction	75
4.5	Cob wall under construction, showing timber formwork in position	75
4.6	Exposed shuttered cob in the wall of a barn in mid-Devon	77
4.7	Upper section of re-built wall under construction using shuttered cob	77
4.8	New wall built using shuttered cob	78
4.9	Detail of window opening in cob wall	79
4.10	Detail of first floor joist support plate and corner reinforcement	80
4.11	Wall plate return at gable wall	82
4.12	Outward leaning cob chimney stack and gable wall	83
4.13	Traditional method of rammed earth construction	86
4.14	Rammed earth; full-height, sliding formwork for vertical panels	87
4.15	Typical formwork for manual rammed earth construction	89
4.16	Alternative methods of securing movable formwork	91
4.17	Tampers for rammed earth construction	92
4.18	Building housing in Yemen using compressed soil blocks	95
4.19	North facing wall of an unrendered cob barn near Exeter, Devon	98
5.1	Relationship between density and thermal conductivity in earth walls	115
5.2	Norden Park and Ride reception building; typical external wall section	117
6.1	Corner crack and outward leaning gable wall in a cob barn	135
6.2	Evolution of the cob wall. Medieval and 19th century buildings compared	137
6.3	Removal of render coat from the gable wall of a cob house	141
6.4	Serious collapse of a late seventeenth century former farmhouse	142
6.5	Principal causes of dampness in earth walls	147
6.6	Rising dampness caused by raised ground levels on a sloping site	148
6.7	Rampant and uncontrolled ivy growth in a Listed cob building	149

6.8   Wall of a cob building showing rat and masonry bee damage   149
6.9   Collapsed section of cob wall, showing rat and ivy damage   150
6.10   Cob cottage in Cornwall, showing severe abrasion damage   150
7.1   The reinforcers. expanded metal lathing (eml) and Helifix
       wall ties   166
7.2   Methods of supporting mass earth walls prior to structural
       repair   168
7.3   Types of structural crack stitches and their positioning   172
7.4   Corner stitches, and re-building failed or collapsed sections
       of wall   173
7.5   Shallow reinforced straight stitches for use in slender walls   173
7.6   Some details of earth block crack stitches   174
7.7   Repair of corner crack and severe erosion using cob tiles   176
7.8   Fired clay tile internal corner crack stitch   176
7.9   Repair of severe erosion at base of wall and repair of shallow
       cavities   177
7.10   A to C. Some alternative methods of re-building failed
       earth walls   182
7.11   Failed wall rebuilt in four lifts of mass cob, and cob face
       repairs   184
7.12   Structural repair using cob blocks toothed into existing wall   184
7.13   Major re-building of a failed cob wall, using cob blocks   185
7.14   Alleviation of rising dampness by means of improved drainage   186

Plate section

The following plate section falls between pages 100–1

1   The Great Mosque at Djenné, Mali
2   Fortified earth city of Shibam, Yemen
3   Public toilets and bus shelter at the Eden Project, Cornwall
4   New cob house at Ottery St. Mary, Devon
5   Two-storey cob extension at Down St. Mary, Devon
6   Information centre at Corfe Castle, Dorset
7   Rammed earth wall at the Eden Centre, Cornwall
8   Rammed earth wall at the Centre for Alternative Technology, Wales

Cover picture

Close-up view of a newly built cob wall; the completed building is shown in
Colour Plate 5

Acknowledgements

Special thanks must go to Robert Nother, Architect and Conservation Officer based in Dorset, who contributed almost all of Chapters 1 and 5 and the section on rammed earth in Chapter 4. Grateful thanks also to the following: Peter Walker and David Clark of Bath University, Department of Architecture and Civil Engineering, for technical advice and guidance on rammed earth construction; Gordon Pearson for sharing his extensive knowledge of rammed chalk and chalk cob building; Dirk Bouwens for information on East Anglian clay lump buildings; Kevin Northmore, Engineering Geologist at the British Geological Survey, Nottingham, for information on the behavioural characteristics of chalk as a construction material; Robert Saxton, Geotechnical Engineer, for carrying out compressive strength tests on soil and chalk samples, and for general advice on soil characteristics; Jill and Mike Smallcombe for help with sourcing photographs for the colour plates; David Webb OBE, formerly Building Research Establishment Overseas Division, for information on compressed soil block construction, and Peter Trotman, former Head of Advisory Services at the Building Research Establishment, for valuable help, over a period of years, with problems relating to dampness in buildings and building materials. Two publications, which have been particularly useful, and to which reference has been made in the preparation of this book are: *The Conservation of Clay and Chalk Buildings* by Gordon Pearson (1992) and *Earth Construction Handbook* by Gernot Minke (2000).

All the photographs included in the book are by either L. Keefe or R. Nother except where otherwise noted in captions and titles. All the line drawings are by the author, apart from Figs 1.3, 4.14, 4.15, 4.16 and 5.2, which were contributed by R. Nother. Cover design by Felicity Keefe.

Preface

In response to the renewal of interest in ecologically sound, sustainable building methods, which has developed over the past two decades, various books
on the subject of building with raw, unfired earth have been published. This
book does not presume to replace or supersede any of these publications, most
of which are listed in the bibliography, but rather to complement them by focus-
ing, firstly, on the ways in which traditional building methods can be adapted to
meet contemporary needs and standards and, secondly, on providing practical
solutions to the problems associated with the repair and conservation of
existing earth buildings. The book is, therefore, arranged in two parts.

Part I, Chapter 1, examines, in outline, British earth-building traditions and the regional distribution of earth buildings. In Chapter 2, because it is
considered that a major impediment to the use of raw earth in new building is
a general lack of understanding of the physical and behavioural characteristics
of earth as a construction material, these are explained and discussed, hopefully in fairly simple terms. Chapter 3 is concerned with the mixing, preparation
and modification of soils for use in building, and in Chapter 4 the various construction methods that may be employed are described in some detail.
Chapter 5 is concerned with how mass earth construction relates to Planning
and Building Regulations, and with the application of technical standards and
codes of practice.

In Part II, Chapter 6, deals with the principal causes of failure in
mass earth walls and how to recognise them, while Chapter 7, which is based
largely on experience of actual building failures, includes a diagnostic survey
procedure for earth-walled buildings, and then goes on to describe in detail
some appropriate repair methods and materials.

A glossary has not been included in the book, as most technical
terms are either explained in the text or shown in line drawings. The reader
is therefore referred to the Index, where one may see, for example, that the
terms 'soil' and 'earth', which are generally thought to be synonymous, are
listed separately. In general, throughout the text 'soil' refers to unaltered
material as dug from the ground, while 'earth' is normally used in the context
of reconstituted or modified material used as a building material, as in the
case of rammed or stabilised earth, for example.

Part 1

Earth construction: methods and materials

Introduction: earth as a 'green' building material

The construction and servicing of buildings probably has a greater impact on the global environment than any other human activity. About 30 per cent of the UK's total energy consumption goes into the servicing of domestic buildings, and a further 8 per cent of the total into the construction of buildings. There is a wealth of published material relating to the need to control resource depletion by reducing energy consumption and to the problems associated with the atmospheric pollution caused by industrial production processes (to which the construction industry is a major contributor). This brief introduction is not, therefore, a polemic, intended to convince the sceptical of the merits of green building. Rather, the aim is to explain how the use of unfired earth and other associated natural materials can play a significant role in reducing energy consumption while at the same time having a minimal impact on the natural environment. In order to do this, it may be helpful to first explain some definitions, or eco-jargon, which sometimes cause confusion, and then to show how earth-based materials perform in relation to these various environmental criteria.

Some terms used in the environmental assessment of building materials and construction methods

- **Sustainability.** This is a fairly vague and non-specific term, much used but perhaps not always clearly understood. It refers to all aspects of human activity and their impact on the earth's ecosystem. Central to the concept of sustainability is the idea that the current world population, or at least those in positions of power, have a clear responsibility to safeguard the earth's remaining resources and its natural environment for future generations. All the terms listed and described below stem from, and form part of, this key concept.

- **Energy efficiency.** This relates mainly to the consumption of energy in buildings in use. A building can be designed and built in such a way that the absolute minimum amount of energy is used, throughout its lifetime, to provide a comfortable and healthy environment for its occupants.

- **Embodied energy,** or **energy intensiveness,** is a measure of all the energy consumed in the extraction, processing, manufacture and transportation of a building component or material up to and including the actual construction process. In a conventional building with a notional 100-year lifespan, the embodied energy input would amount to about 10 per cent of the total energy consumed during this period (B. and R. Vale, 1993). However, according to another source (Borer and Harris, 1998), 'As new buildings are designed to be more energy-efficient, the energy-in-use portion [referred to above] will fall. . . . The embodied energy factor will then become relatively more significant and could constitute up to 50 per cent of the total energy use of an energy-efficient building over 30 years.'

- **Zero emissions.** The concept of zero emissions is based on the use of natural materials, employed in such a way that no carbon dioxides or other potentially harmful chemicals are released into the atmosphere. Under this general heading may be included the process known as 'out-gassing' or 'off-gassing'. This refers to the use in contemporary buildings of synthetic materials, which release toxic, sometimes potentially carcinogenic gases from volatile organic compounds and solvents into the interior of a building. Because many modern buildings are of non-breathable construction and lack effective ventilation, the emission of these gases can result in what is known as SBS or Sick Building Syndrome, the potentially damaging effects of which are now fairly well known and understood.

- **Natural materials.** These are building materials that may be either organic or mineral in origin, either grown in the earth or extracted from it, which are used in unaltered form. This would include, for example, organic materials such as timber, straw and other plant fibres, and animal hair; also earth minerals such as stone, sand, gravel and subsoil. Unlike manufactured materials, natural materials are not processed or modified in a way that requires the consumption of significant amounts of energy.

- **Renewable resources.** In terms of energy consumption, resources used in the production of buildings may be classified as either renewable or finite. Obvious examples of finite materials are fossil fuels such as coal, natural gas and oil; also rocks containing metalliferous and other mineral ores. Most natural materials are renewable except in cases where they undergo irreversible change through industrial processing, as a result of which they become either non-biodegradable or incapable of recycling.

- **Recycling.** When a building becomes redundant or, as is often the case, the site it occupies is required for redevelopment, it will be demolished. Clearly, in environmental terms, to demolish a sound, re-usable building for no better reason than that of economic expediency is very wasteful. Although it is possible to recycle some building materials, especially those from nineteenth- or early twentieth-century buildings, the high labour costs of doing this are such that potentially valuable materials often end up in landfill sites, and recycling of cement-based materials such as reinforced concrete, and bricks which have been bedded in strong cement/sand mortars, is virtually impossible. Unbaked earth and other natural organic materials are, by contrast, eminently recyclable. The issues of finding suitable uses for redundant earth buildings and of their economic repair and conservation are discussed in the introduction to Part II of this book.

- **Waste materials and their use in building.** In a consumerist, throwaway society, where built-in obsolescence is considered normal, the disposal of vast quantities of 'waste' material poses a considerable, apparently insoluble, problem. All the more reason, therefore, to take recycling seriously. Waste materials which have traditionally been used in conjunction with earth for building construction include straw, flax, hemp and other organic fibres; also sawmill waste – sawdust and wood shavings. Currently, research is being carried out to assess the potential of other waste materials, including ground glass aggregate and shredded rubber tyres, as elements in mass earth walls (University of Bath, 2002–4).

- It should be noted that the current **Building Regulations for England and Wales**, unlike those of some other European countries – Germany and Denmark, for example – do not take into account either embodied energy or recyclability in the energy assessment of buildings. This issue is further discussed in Chapter 5.

The performance of earth-based materials

High thermal mass external walls in domestic buildings are considered to be more energy efficient than walls of lightweight construction, mainly on account of their high thermal capacity; that is, their ability to store heat and then to release it slowly when the heat source is removed, in much the same way as a night storage heater. Earth is ideally suited to this form of construction because, although its density is lower than that of concrete blockwork, its thermal characteristics are not dissimilar to those of fired bricks, sometimes marginally better. Earth may be used to construct solid, cavity or cellular walls, and may be mixed with lightweight aggregates to improve its thermal performance (see Chapter 5).

The concept of the embodied energy of building materials is one that is easily understood. However, to calculate precisely the overall amount of energy consumed, from extraction through to construction for a given building material, is a very complex process. The data shown in Table I.1, which has been compiled from various published sources, must therefore be regarded as providing, at best, a rough approximation of energy costs in kilowatt-hours per cubic metre (kWh/m^3) (energy required to produce one cubic metre of material).

It is not clear to what extent the data shown in Table I.1 take into account the energy costs of transportation. Energy consumed in transporting

Table I.1 Embodied energy consumption of selected building materials

Building material	Energy consumption (kWh/m^3)
Cement (OPC)	2,640*
Fired brick (solid)	1,140
Chipboard	1,100
Lime	900*
Plasterboard	900
Concrete block	600–800
Fired brick (perforated)	590
Calcium silicate brick	350
Natural sand/aggregate	45
Earth	5–10
Straw (baled)	4.5

Data from various sources, including Intermediate Technology Development Group and the Centre for Alternative Technology.

*Dry bulk density of OPC is $1,200kg/m^3$ and that of hydrated lime $600kg/m^3$.

materials over long distances can significantly increase their embodied energy input. For example, it has been estimated that if a truck travels a distance of 150 km (93 miles) carrying a load of fired bricks and returns empty, this would increase the embodied energy costs of the bricks by 40 to 50 per cent (B. and R. Vale, 2000). The same applies to softwood timber imported, as it mostly is, from Scandinavia or North America, where it is estimated that the embodied energy of locally sourced timber is only about 15 per cent of that of imported material (Talbott, 1995). Lime and cement are included in Table I.1 for comparative purposes and because the addition of small quantities of lime to earth renders may be justified in certain cases (see Chapter 4). It may be noted that the energy consumed in cement production is about three times that required to produce a similar quantity of hydrated lime. Furthermore, it has been estimated that the manufacture of cement, which is a complex, highly industrialised process, is responsible for 10 per cent of global CO_2 emissions. Lime, on the other hand, can be produced locally using fairly simple traditional technology (Wingate, 1985).

There are many hundreds of surviving limekilns in Britain; an indication of how widespread lime production was in former times, before the advent of industrially produced hydraulic lime and Portland cement from the mid-nineteenth century onwards. Reference to Table I.1 will show that even concrete blocks use around 100 times the energy of that required for earth construction, and that the production of straw, which becomes a waste material once the wheat or barley grains have been separated out, consumes only negligible amounts of energy.

Zero/low emissions. As noted above, in the case of earth, straw and other natural organic products, the amounts of CO_2 and other chemicals that they release into the atmosphere are so minute that they may be regarded, relatively speaking, as zero emission materials. With regard to internal emissions, or out-gassing, earth and unprocessed organic fibrous materials are entirely non-toxic and therefore present no risk to health, provided they remain dry. The one exception to this may be the presence of radon gas, which is normally found only in subsoils overlying granite rock formations. Lime, although produced by the high temperature burning of chalk or limestone, is also non-toxic and quite harmless once it has carbonated. In fact, in former times, limewash was valued for its mild antiseptic and biocidal properties, and for this reason was applied to the internal walls of agricultural buildings in which animals were housed.

Renewability and recycling. Earth, although not renewable in the same way as organic materials such as timber and straw is, over very long periods of (geological) time, being constantly created through the weathering of rock. It is a material that is universally available, often at little or no cost, because in many civil engineering and housing development projects thousands of tonnes of unwanted subsoil are excavated, which are then transported, at great cost, to landfill sites that may be located many miles from the construction site. Tristram Risdon, in his *Survey of the County of Devon* written in the early seventeenth century, describes the process of cob walling, which he concludes by noting that: 'when any such Walls are pulled down to be re-built, they commonly make fresh Cob with other Earth; the Value of the old as Manure for land sufficiently compensating the Cost of the new' – an early example of recycling, which has as much relevance now as it did four centuries ago.

Chapter 1

Earth building in Britain – international context and historical background

By R. Nother, with an introduction by L. Keefe

British earth building in an international context: a brief history

It is estimated that at least 30 per cent of the world's population, some 1.5 billion people, live in houses constructed of raw (unfired) earth. Earth build- ings exist in every habitable continent but are far more numerous in hot, arid regions of low rainfall where there is a dearth of timber suitable for building. Archaeological and historical research would suggest that earth has been used for building construction for about 10,000 years, the earliest surviving remains having been found in the Middle East. By the end of the pre-Christian era, earth building had spread throughout all the world's ancient civilisations, from China through to North Africa and the eastern Mediterranean, and including Central and parts of South America. The earliest recorded building technique is 'adobe', unfired mud bricks containing straw. Later, mass earth walls were constructed of earth rammed into timber formwork, and 'piled earth' – wet earth with fibre reinforcement, made into balls, which are then thrown up on to the wall head and beaten into place. This technique – which is not dissimilar to that of British cob building – is used in conjunction with adobe bricks to construct large, impressive buildings of up to 10 storeys with walls 30 metres high, in the Middle East and North Africa to this day – an unbroken, centuries-old tradition (see Colour Plates 1 and 2). It is estimated that in China around 45 million people live in houses constructed of either rammed earth or mud brick made mainly from loess – deep deposits of wind-blown silt laid down during the Pleistocene period (Houben and Guillaud, 1994). When the Spanish and Portuguese arrived in Central and South America, not only would they have brought with them their own, Moorish-inspired, knowledge of earth building, but they would also have found a well-established indigenous earth-building tradition. As a result, numerous large structures, including cathedrals, churches and other public buildings, as well as houses, most of which were built with adobe bricks, survive from the Spanish colonial period.

With regard to Europe, although there is evidence of the wide-spread use of earth construction in the Mediterranean regions in pre-Christian and Roman times, in north-western Europe the picture is rather less clear. There is some evidence that, during the Roman occupation of Britain, rammed and 'puddled' earth was used in the construction of military and other build-ings, notable examples having been found at Norwich Castle and Verulamium (McCann, 1995) but, not surprisingly perhaps, little else has survived, although evidence has been found of the making and use of unfired earth bricks in Leicestershire (Harrison, 1984). After the Romans departed from southern Britain, having failed either to subdue the Scots or to occupy Ireland, their build-ing technology seems to have been largely forgotten, and for about six cen-turies building construction remained at a fairly basic level – although carpentry became quite sophisticated towards the end of the period. The walls of all but the highest-status buildings, churches and manor houses, for example, were usually timber framed with an infill of wattle-and-daub or similar earth- or chalk-based materials. The same was probably also true of most of northern Europe, Scandinavia and northern Germany, for example, where much of the Anglo-Saxon British population originated. The earliest examples of mass earth walling in Britain date from the thirteenth century. However, many of these earlier cob or 'mud' buildings must be regarded as transitional in the sense that their walls were not fully load-bearing. Their main roof trusses, or principal rafters, rather than bearing directly on the earth wall head, formed the upper part of a 'cruck' frame, which was either supported on the stone-pinning course or built directly off the foundation. A variation of this system, found mainly in south-west England from the fifteenth century onwards, was the 'jointed cruck' in which the feet of the principals were jointed into curved full or half-height wall posts. Fully load-bearing earth walls became increasingly common from the late sixteenth century onwards and within a hundred years or so came to be almost universally adopted (see Figs 1.1A to C).

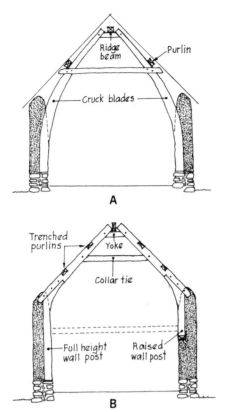

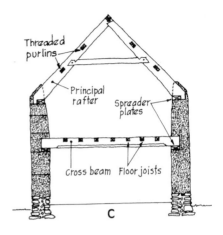

1.1

Typical constructional details of some early cob and mudwall domestic buildings: (A) fourteenth-century cruck frame, (B) early sixteenth century – two types of jointed cruck and (C) late sixteenth and early seventeenth century – two storey with A-frame roof trusses

Note: common rafters (not shown) are supported on purlins and at the wall head

The earth-building revival

One result of the industrialisation of the Western World that took place during the nineteenth and early twentieth centuries was that many traditional building methods and materials were swept aside by a wave of new technology based upon industrial-scale mass production and improved transport networks. As a result many traditional building techniques fell into disuse and the craft skills associated with their use were largely forgotten. Apart from some experimental building projects during the decade following the end of World War I, the renewal of interest in raw earth as a building material dates from the second half of the last century, largely as a result of international crises and increasing concern over the need to improve the housing conditions of an ever-expanding 'third world' population. In the chaotic conditions that existed throughout much of Europe following the end of World War II, there was an urgent need to provide shelter for hundreds of thousands of homeless people; in Germany and eastern Europe, where there was an acute shortage of building materials and all industrial production had virtually ceased, some government agencies responded by building thousands of houses and entire settlements using rammed earth or mud bricks. However, during the period of economic recovery that took place during the following two decades, interest in alternative building technology waned; that is, until the 'energy crisis' of the early 1970s, which provided the impetus for consideration of the need to significantly reduce global energy consumption. Meanwhile, from the 1960s onwards, overseas aid programmes initiated and funded by the UN and developed nations, including the USA, Britain and France, were researching and actively promoting the use of appropriate technology building, including raw earth construction, in the poorer, less developed parts of the world. At the same time, in Australia, the USA and, later, in France, raw earth was being used successfully for the construction of new housing for the domestic market. In this respect Britain has lagged far behind other Western nations. Other than the valuable work of the Overseas Division of the Building Research Establishment (whose activities were curtailed through withdrawal of government funding in the mid-1990s), some academic research, and building projects initiated by private individuals and organisations such as the Intermediate Technology Development Group and the Centre for Alternative Technology, not a great deal has been achieved. However, hopefully all this may be about to change.

Earth buildings in Britain – types and distribution

In Britain, the greatest concentration of surviving earth-walled buildings is in south-west England. However, other areas of Britain are not without significant numbers, and this chapter sets out to show where such buildings can be seen, the materials commonly used and their main constructional forms. It will also examine the range of buildings in terms of original use and status. Many areas have surviving examples predominantly of one wall constructional type, while in others a number of different types co-exist. There are potential pitfalls in placing the types of earth-based building materials and associated constructional techniques in clearly defined categories as some individual buildings embrace more than one walling type, and in some areas the constructional technique for virtually the same material varies. This, perhaps, is what might be expected in processes deriving from a variety of local vernacular traditions. Nevertheless, classification into broad categories helps to convey the essential qualities of the material. Following an account of these categories is an indication of the known existing spread of earth buildings throughout Britain and the predominant types found in each area. In all probability, actual numbers of surviving earth walls far exceed those identified here, but many of these are not immediately recognisable due to being cloaked in later fabric or absorbed within subsequent modifications.

The intention is to focus on buildings where subsoil earth, usually containing clay, chalk or a combination of these, has been used to construct load-bearing walls. Hence turf, being a topsoil material, is not included, although some turf-walled buildings survive in Ireland and Scotland. Also excluded are such techniques as wattle and daub, where the earth content is seen as part of a composite infill panel within a framed structure and which does not contribute significantly to the load-bearing function of the frame. In medieval times mass earth walls were referred to almost universally as 'mud-walle', a term no doubt acceptable at the time but that might these days be considered somewhat pejorative. The use of the word 'cob' is first recorded in Devon and Cornwall at the beginning of the seventeenth century, and over the years has become a generic term used to describe all wet mixed and placed, fibre-reinforced mass subsoil walling techniques, although some regional names still remain in use. For example, in Scotland the term 'mudwall' is still commonly used and in Wales the technique is known as 'clom'. For convenience, and in order to avoid confusion, the terms 'cob' or 'mudwall', which may be regarded as synonymous, will be used throughout the text to describe this building method. The majority of surviving buildings with walls of this type were built between the early seventeenth century and the mid-nineteenth century. Although there are substantial numbers surviving from the sixteenth century and as far back as the fourteenth century, very early buildings tend to be few in number.

Earth-walling types

For convenience, the categories set out below are presented as individual types, whereas in reality it is not uncommon to find combinations of them within the same building. For instance, examples can be found of a wet mix laid *in situ* combined with clay lump, or with a form akin to mud-and-stud. For the most part, the building methods described are longstanding and traditional, apart from rammed earth and clay lump, both of which are non-indigenous techniques introduced into Britain in the 1790s by agricultural 'improvers', social reformers, the owners of large estates and their architects.

Commonly, all the types of earth walling described here have a supporting base of a different material, termed an underpin course. In early examples, the underpin course is constructed of stone gathered from the locality around the building. During the late eighteenth and nineteenth centuries brickwork became increasingly common for this purpose. Examples survive without an underpin course, although these are relatively few in number. Where no underpinning course is present it is likely that the building had very humble origins, perhaps a tenant farmer's outbuilding or a squatter's cottage. The inclusion of the underpin course is considered good practice because it raises the earth walling above ground level and provides some protection from the adverse effects of groundwater, surface water and splashback. Also common to most traditional examples originally was the thatched roof. With its generous overhang, this gave good protection to the head of the earth walling. When coupled with a substantial plinth and, sometimes, an external coating to the wall surface, this provided the prospect of reasonable durability.

Wet mix laid *in situ* (cob and mudwall)

This is by far the most widely occurring type, and has been described as a primitive form of concrete, the material forming a monolithic mass. However, this should not be confused with the modern form of concrete, which relies on Portland cement to form a chemical set, and is not reversible. As with all traditional forms of earth walling, this type gains its cohesiveness mainly through the cementing effect of the clay binder, with organic fibre reinforcement also playing an important role. The process is reversible through the addition of water. Essentially, the material comprises clayey subsoil and aggregate, or chalk, usually with the addition of such material as straw or heather, well mixed together. Whilst in many areas clay formed a key ingredient, in some localities, notably parts of Dorset, Hampshire and Wiltshire, the material is comprised almost entirely of chalk. In these cases, the chalk is in aggregate form, and is not to be confused with chalk block, or 'clunch', which in essence is a form of masonry construction.

Common to all the different mixes was the presence of just enough water to make the mix suitably workable during the construction process, having a consistency similar to stiff dough or putty. In most cob mixes the addition of some form of organic fibrous material was, and still is, considered to be essential, though in some areas this was apparently not the case. For example, in Northamptonshire the local ironstone clay is usually stable enough without the addition of straw. In parts of Buckinghamshire a calcareous clay overlying the Portland beds is used, producing a mix which can be particularly strong and durable if kept reasonably dry. This is known as 'witchert', derived from the term 'white earth'.

The constructional techniques associated with cob, or mudwall, vary with place and time. The predominant method involves a series of unshuttered lifts, each lift being within the region of 300 to 600mm. An alternative process using similar materials but with lifts of only about 50 to 150mm is found in the Solway Plain area. Professor Ronald Brunskill has termed these respectively the 'slow process' and the 'quick process' (Brunskill, 1987). Of key significance is that the quick process can be continuous through to the completion of the wall, while the slow process requires a time lapse between lifts. In the quick process, it is usual to find a bed of straw laid between lifts, perhaps to assist in the drying-out process through capillary action. A building recently constructed using the quick process is shown in Fig. 1.2. The entire building was constructed, using volunteer labour, between sunrise and sunset on midsummer day, 21 June 1997, on the Beaulieu Estate in the New Forest area of Hampshire (Countryside Education Trust, 1997).

Often, the time lapse between lifts in the slow process is in the region of seven days, but varies with weather conditions, time of year and content of the mix. Commonly, external walls of this type of early date are about 600mm thick and sometimes much more, while internal and boundary walls are often somewhat thinner. Later, eighteenth- and nineteenth-century cob walls in Devon tend to be 500 to 550mm thick, but only 400 to 450mm in agricultural buildings. In some areas, chalk and mud walls have been constructed using shuttering, although this is not immediately apparent unless some of the rendering has fallen away or unrendered areas still display evidence of board marks or board junction lines.

This alternative form, which is sometimes termed 'puddled clay', used, as its name suggests, a wetter mix than for those walls laid unshuttered. A blend of clay, silt, sand and gravel was mixed with straw and water, and thrown into shuttering before being lightly tamped (rather than rammed or trodden) and then allowed to dry, a process that might take many days or even weeks. It is believed that shuttering was used increasingly through the late eighteenth and nineteenth centuries, enabling the overall thickness of the wall to be reduced and speeding up construction. Details of how these mass earth walls were, and still are, constructed can be found in Chapters 3 and 4.

1.2
Reconstruction of a seventeenth-century New Forest squatter's cottage, built in one day using the 'quick process'

Mud and stud

This technique is associated mainly with one area of Britain, namely Lincolnshire, but is also found in parts of Scotland. The inclusion of this technique here is considered appropriate because the earth contributes to the structural performance of the overall building, unlike wattle and daub. Although these walls contain a timber framework, which has a major load-bearing function, this is totally covered by an applied coating of earth, resulting in an external appearance very similar to that of cob. Indeed, the mix comprises ingredients very similar to that of other mass *in situ* earth-walling types, the major difference being that the total wall thickness is in the region of only 200 to 300mm. Lifts are normally about 500mm.

The upright posts of the frame, commonly between two and three metres tall and about two metres apart, traditionally were earth-fast, but more recent examples have their posts supported on pad-stones. These usually had a morticed joint with the wall plate. Vertical laths, usually of riven ash, are nailed to the bottom plate (or, more usually, driven into the underpin course), horizontal mid-rail and head. Unlike wattle and daub, there is no horizontal interweaving between the vertical laths. Instead, the structural integrity is achieved through the applied earth being considerably thicker than that for wattle and daub (see Fig. 1.3).

Clay lump

In many ways the ingredients of clay lump resemble those described previously. However, unlike the former materials, which often require some modification to achieve the appropriate balance of ingredients, clay lump is usually used as dug from the ground. The key difference, though, is that the material is pre-cast in block form, the sizes normally ranging between 100 × 75 × 225mm and 225 × 225 × 450mm. This results in most of the drying shrinkage taking place before the construction of the building. The blocks usually were laid in a mud mortar, or sometimes a lime-based mortar. In Britain, the material is associated principally with the heavy boulder clays of East Anglia, which are prone to considerable drying shrinkage, and which generally form a high proportion of the overall mix of ingredients. The pre-cast nature of the material allows more precision in assembly, and this in turn enables the walls to be thinner. Generally, external walls are about 250mm thick, and internal partitions about 190mm (Fig. 1.4).

Rammed earth

Rammed earth, or *pisé de terre*, as it is known in France, never really caught on in Britain, despite strenuous efforts to promote its use at the beginning of the nineteenth century. For example, one mid-nineteenth-century correspondent from Sidmouth in east Devon had this to say concerning *pisé*:

1.3

**Details of mud-
and-stud
construction**

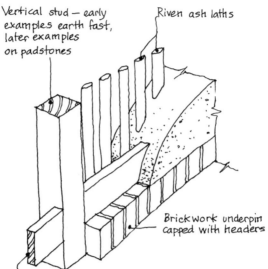

Vertical stud — early examples earth fast, later examples on padstones

Riven ash laths

Brickwork underpin capped with headers

Sole plate (relatively rare — the lower laths were normally jammed into the joints of the brick underpin)

Note: In England, this technique is confined largely to parts of Lincolnshire and the fringes of some adjacent counties, although examples have been identified in the Lancashire coastal region and in gable construction in Cumbria

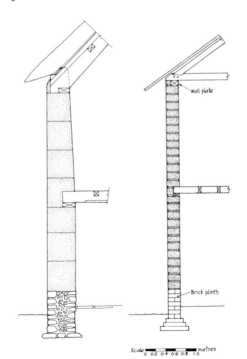

wall plate

Brick plinth

Scale ▬▬▬ metres
0 0·2 0·4 0·6 0·8 1·0

1.4

**Cob and clay-lump
walls compared**

> This process has been introduced sparingly into the west of England. When I was a boy, I recollect witnessing the erection of two or three houses in this way; but it was looked upon as a novelty. It was done by ramming earth in between two planks with rammers. The earth was dry, having only the ordinary dampness of the ground, and without straw.
>
> (Hutchinson, 1857)

Clearly, in this case the innate conservatism of the West Country builders prevailed. However, the technique was successfully adapted for use with chalk, mainly in Hampshire, from about 1830 onwards. Numerous buildings, including some very grand houses, were constructed using this method, at least 100 of which are still standing. Rammed earth construction methods, both ancient and modern, are described in detail in Chapters 3 and 4.

These, then, are the principal traditional types of earth walling which can still be found in Britain. Of these, it will be seen that cob, and its variants, is by far the most common, and archaeological evidence shows that it is a very long-established technique. Among the earliest examples for which there is clearly recorded evidence is the mud walling revealed during excavations at Wallingford Castle in Oxfordshire, dating from the early thirteenth century. This early use is supported elsewhere by documentary evidence. Salzman (1952), for example, draws attention to manorial accounts, which refer to an earthen wall at Southampton in 1312. The earliest known cob building in Devon has been dated, by means of dendrochronology, to 1299/1300. In order to improve their weather resistance and perhaps also their appearance, external render coatings, internal plasters and paint finishes were applied to earth walls, particularly those of domestic buildings, from the late eighteenth century onwards. Finishes for earth walls are discussed in detail in Chapter 4.

The distribution of earth buildings in Britain

Having briefly described the main types of earth-walling construction, we now take a look at where in Britain they can be found. Having first looked at Ireland, the examination returns to mainland Britain, and, starting with Wales, proceeds clockwise and ends in south-west England where earth buildings are most numerous. The map at Fig. 1.5 shows the areas within which significant numbers of earth buildings are known to have survived.

Ireland

Most earth-walled buildings in Ireland tend to be fairly modest in scale, having been built by, and for, a predominantly rural, agrarian population – mainly

1.5

**Regional
distribution of
surviving earth
buildings in Great
Britain and Ireland**

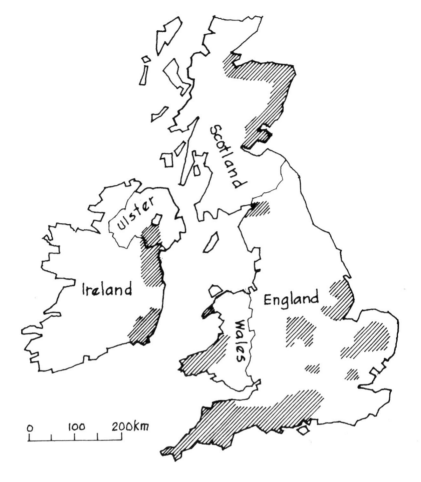

smallholders and landless labourers. Ireland has a tradition of turf-walled houses, and indeed these might be regarded as inferior as they were prone to fairly rapid disintegration and required frequent repair or even rebuilding. Examples of most of the techniques described previously can be found in Ireland. However, the majority of the surviving examples may be classified as cob or mudwall. Most of these were built within the period 1500 to 1900. While there is evidence of earlier examples, it is believed that none survive to the present day. Mudwall, or 'tempered clay', was considered more satisfactory than turf walling, and many houses constructed of this material remain inhabited. Documentary evidence indicates that buildings of such material were not confined to rural areas, although the number of urban examples surviving today is probably very few.

Whilst such buildings were at one time spread profusely throughout much of Ireland, today this has for the most part been reduced to a scattering of rural buildings with significant numbers surviving in only two areas. One area is within the eastern half of the country between the south-ern shores of Lough Neagh and around Dublin, extending to the eastern coast. The other area is in the south-east, extending westwards from Wexford almost to Waterford and northwards to around Arklow. These are the areas that contain the most extensive deposits of fine glacial clays. Most of the surviving buildings are modest dwellings or farm buildings, the major-ity being single storey, although a few two-storey examples containing substantial amounts of earth walling survive. Most of the surviving examples were built with unshuttered walling, but there are some situations where shuttering was known to be used. In some cases horizontal timbers were built into the earth walling as it was raised, providing a form of reinforcing. Predominantly, the buildings are long and low, originally having thatched roofs. Generally, roof construction consisted of cruck frames or simple A-frames. Wall plates are not common, it being more usual to place a flat stone under any rafter or joist that was built into or supported on the earth wall.

1.6
Eighteenth-century 'clom' cottage, re-erected at the Museum of Welsh Life, St Fagans, Cardiff

Wales

Here, the material used predominantly is known locally as 'clom'. It was laid *in situ* following the 'slow process', albeit that in some walls the layers are only 100 to 200mm deep. Clom buildings were associated with poorer rural dwellings, usually having been self-built by their occupiers, and there are around 250 known surviving examples, mostly dwellings and small agricul-tural buildings. The majority of these are located in South Cardiganshire, North Carmarthenshire, and the Lleyn Peninsula. Those in the south-west com-monly used scarfed or jointed crucks, while those in the Lleyn Peninsula had simple A-frames bearing on the tops of the walls. A typical example would be a two-roomed single-storey dwelling with straw thatched roof and a smoke hood/chimney in wicker with an earth daub or lath-and-plaster (Fig. 1. 6). Many are known to have had wicker or wooden chimneys. Sometimes, one end is lofted, this being known as a 'croglofft'. The walls were raised separately, not bonded together, to a height of 2.0 to 2.7 metres on a stone underpin layer. Openings were minimal in number and size, some buildings originally being windowless. Most of the Lleyn cottages measure about 7.6 by 4.6 metres on plan. A few examples include the use of large earth blocks, particularly in the construction of a chimney breast. A signifi-cant growth in the numbers of such cottages occurred in the late seventeenth and early eighteenth centuries, when the legal enclosure of common land was at its most rapid. This suggests that many originated as squatters' cottages, scattered rather than grouped, at the fringes of unenclosed commons. Sadly, many of the surviving examples are now ruinous, and are likely to be lost in the near future.

1.7
Clay Dabbins. Farmhouse and restored barn at Burgh by Sands, Cumbria

North-west England/South-west Scotland

Whilst probably there are scattered surviving examples of earth buildings throughout north-west England, including a number in the Fylde area of Lancashire, it is in Cumbria, around Carlisle and on the Solway Plain, where the greatest concentration in the region, probably several hundred buildings, can be found. The earth-walled buildings here are known locally as 'clay dabbins' or 'clay daubins'. They are built of unshuttered cob, using the 'quick process' described previously. Originally the spread of these buildings extended from the Solway to Berwick, being abundant on both sides of the border. However, only a small number survives on the Scottish side. The earliest surviving examples were formerly thought to be of seventeenth-century date, these being usually single-storey, sometimes with an attic, and having a cross-passage alongside the main hearth or inglenook. Most of these early examples have cruck frames, and the clay wall is constructed on top of a boulder and cobble plinth. Subsequent research using more sophisticated dating technology, however, suggests that some of these buildings may be as early as the fourteenth or fifteenth centuries.

Many of the later examples are of two storeys, and use simple trusses rather than crucks. The earliest roofing materials were turf, straw, heather or bracken. Many buildings still have thatched roofs, but there has been a tendency to replace earlier coverings with sandstone flags or slate. In the use of the 'quick process', documentary evidence supports the view that in low buildings the walling could be constructed within one or two days. The development of a quick process, enabling subsequent speedy rebuilding, possibly stemmed from the needs of a population suffering a turbulent existence in the Borders during the Middle Ages and later. Whatever the reason for such a technique, its use continued through to the mid-nineteenth century, although interspersed with examples using the slow process. A typical early clay dabbins building is shown in Fig. 1.7.

North and east Scotland

Whilst earth buildings survive in the Highlands and Islands of Scotland, these are few in number and mostly of turf construction, and there is very little extant evidence of subsoil earth walling in these areas. The greatest number of surviving Scottish earth buildings are close to the east coast, extending from the area around the Moray Firth south and eastwards to the peninsula between the Firths of Tay and Forth. Field-study data indicate that all the types of traditional earth walling described previously are to be found, with the possible exception of clay lump. An additional form of construction, largely confined to parishes around the Spey estuary, is 'clay and bool'. This involved building up the wall in lifts of large beach stones or field pickings placed against shuttering, with a clay mix packed in between and around them. It may be seen as a hybrid form, perhaps more closely related to masonry. In regions where monolithic earth has been used, without the presence of large boulders, commonly this has been shuttered during construction.

Many of the earlier buildings, some surviving examples of which date from the eighteenth century, are single storey. As with the Welsh and Solway Plain examples, most were originally thatched. A greater proportion of the later buildings have two storeys. Those with thinner and more precise walling through the use of shuttering, coupled with larger windows and roof coverings other than thatch, at first sight are difficult to identify as being of earth construction. Whilst actual numbers surviving in Scotland cannot be given here, it is known that there are around 16 in Moray district which largely retain their original form.

Lincolnshire

Mud-and-stud constitutes the predominant form of earth building in Lincolnshire and around 350 buildings of this type are known to exist, many of these being centred on the district of East Lindsey. Of these, about 20 remain fairly close to their original condition. Documentary evidence shows that mud-and-stud was in use in the mid-seventeenth century; for example, a parish overseer's account, which appears to refer to a mud-and-stud building erected at Saleby in 1686, has been recorded (Machin, 1997). Thousands of earth buildings in the area were built in this way, including cottages, farmhouses, barns and almshouses. However, many were lost in the early part of the twentieth century due to their perceived insanitary and degrading qualities. Almost all examples conform to a near-standard pattern, being essentially of one storey and attic, and having a lobby entry in front of a central brick stack, with a room to either side (Fig. 1.8).

1.8
Mud-and-stud houses at Thimbleby, Lincolnshire

East Midlands

In this area, the predominant earth-building method was unshuttered cob, known locally simply as 'mud'. Here, it is very much associated with humble dwellings, farm ancillary buildings and boundary walling. As in many areas, there is no certainty about how many such buildings survive today. In the Harborough district of Leicestershire, there are about five statutorily listed structures known to contain mud walling. Of these, one is a cottage, and the others outbuildings and boundary walls. The actual number surviving is believed to far exceed those identified in the statutory listings. In Northamptonshire, very nearly 100 statutorily listed buildings are described as containing earth walling, although, again, there are believed to be many more. Virtually all of these are in Daventry district. Of those identified, about 35 are dwellings, and the remainder barns, outbuildings, dovecotes and boundary walls. Few walls exceed three metres in height, and most buildings are of humble scale socially and physically. Roofing originally was thatch. Most surviving examples are of eighteenth- and early nineteenth-century date. Throughout the region, it is believed that about 1,000 or so buildings remain.

East Anglia

East Anglia is second only to the south-west region in terms of surviving numbers of earth buildings. This area yields a chalky boulder clay which is well suited for earth walling, albeit that in its unmodified composition it is prone to considerable drying shrinkage. Hence, clay lump, or clay bat, became the predominant form of earth building where this material was available. This was because the prefabrication of the clay blocks enabled most of the drying shrinkage to take place prior to the blocks being laid in the walls. Buildings of such construction can be found in south Norfolk, north Suffolk, Cambridgeshire and Essex. The majority are two-storey houses, in terraces, pairs and detached. Some of these are relatively large and of high status (Fig. 1.9). Also, there are numerous other building types using clay lump, including several large barns and other agricultural buildings. Large numbers of these clay-lump buildings can be attributed to the early and mid-nineteenth century, and the technique continued in common use until the 1930s, when it was used to construct, *inter alia*, local authority housing. During this period, shuttered clay (or puddled clay) construction was also in fairly common use. It is estimated that there may be up to 20,000 surviving earth buildings, mostly of clay-lump construction, in the East Anglian region (Bouwens, pers.com).

Whether the clay-lump technique had an earlier presence in the locality is open to question, but in all probability *in situ* mass earth walling was abundant in earlier centuries. This and other types of earth walling survive, interspersed among the more profuse clay-lump examples. Clay-lump buildings are not always immediately recognisable, due to their relatively

1.9
**Clay-lump
building, formerly
a home for single
mothers, at
Bressingham,
Norfolk, *c.* 1850**
Photo: Dirk
Bouwens

slender walls and the tendency, in later buildings, to apply an external half-brick or flint facing. It was not unusual for clay-lump walls, particularly those of agricultural buildings, to receive a coating of coal, tar and sand externally. Most clay-lump buildings have an underpin course of bricks, and hence clay lumps generally occur in sizes having a modular relationship with that of bricks. Many are about two bricks long and one brick wide, and commonly the equivalent of one or two brick courses high. The flint or brick plinths are normally between 300 and 1,200mm high (Fig. 1.10).

Buckinghamshire and Oxfordshire

In contrast to the mud walling of the East Midlands, the chalk and Portland limestone-based material known as 'witchert', found in Buckinghamshire in the vicinity of Haddenham, is held in high regard. Here, the material as dug usually required little adjustment to make it suitable for building. It was raised in unshuttered lifts, and buildings of this material survive from the seventeenth century in great numbers, examples continuing to be built until about 1900. The fine quality of the material, which has been shown by analysis to contain around 45 per cent calcium carbonate and 15 per cent expansive clay, is reflected in the range of buildings in which it was used. In addition to fairly humble dwellings, barns and boundary walling, it is found in imposing villas and chapels. Some walls exceed six metres in height without intermediate

lateral restraint. However, recent experience has shown that witchert is just as susceptible to the effects of excess moisture as most other soils. Around 300 surviving examples have been identified, about 100 of which are in the village of Haddenham (see Fig. 1.11). A number of earth buildings occur close to the border between Oxfordshire and Buckinghamshire, and these are believed to be of witchert. Eleven buildings in south Oxfordshire have statutory listing descriptions that refer to earth walling, but it is thought that the total number far exceeds this.

South and south-west England

South and south-west England undoubtedly has greater numbers of earth buildings than anywhere else in Britain. Within this area, parts of Hampshire, Dorset and Wiltshire have substantial numbers, but it is in Devon where by far the largest numbers survive. Most categories described previously can be found in the south and south-west, but wet earth laid *in situ* predominates. Also, rammed earth buildings in significant numbers have been identified, most of these being in the Winchester and Andover areas of Hampshire, where the chalk is particularly suitable for this technique. Of those surviving today, most were built between the sixteenth and nineteenth centuries. The tradition did not die out entirely during the later nineteenth century, in spite of easier distribution of cheap building materials with the coming of the railways. Very few new buildings were erected in the early twentieth century, some rare examples being the experimental earth dwellings built at Amesbury, Wiltshire, in 1919–20 to W. R. Jaggard's designs (see Fig. 1.12) and a number of others built some years later by Jessica Albery and B. H. Nixon (Pearson, 1992).

1.11
Witchert houses and boundary walls at Haddenham, Buckinghamshire

1.12
Chalk cob house at Amesbury, Wiltshire, 1920. Note lift lines in gable end wall

Whilst earth walling in this region is generally associated with relatively humble rural dwellings, barns, outbuildings and boundary walling, this is by no means exclusively so. Some early dwellings were relatively high on the social scale.

Bowhill, Exeter, Devon, a mansion of the late fifteenth century, exemplifies this, as does Hayes Barton, East Budleigh, the birthplace of Sir Walter Raleigh, also in Devon, and Cruck Cottage, Briantspuddle, Dorset, of similar date to Bowhill, and originally a hall-house built probably for a yeoman farmer. Also, during the early nineteenth century, imposing dwellings – town houses and marine villas – having earth walls were built. Some of these were urban, as at Dawlish, Sidmouth and Teignmouth in Devon (Fig. 1.13) and some rural, as at East Morden, Dorset. It has been estimated that in Devon alone there may be 20,000 houses and an approximately equal number of barns and other outbuildings, together with numerous boundary walls.

It is claimed that below the 'Middle Manor House' size, cob becomes as common as stone for farmhouses in Devon and probably the dominant material in the county for cottages. The term 'cob' is used generically throughout this region, the unshuttered wet mix laid *in situ* being widespread and abundant. However, the content of the mix varies from place to place, ranging from the almost pure chalk in the east (Fig. 1.14) through the clay and aggregate of central Devon's Culm Measures and Permian Breccias, to the 'clob', with its shilf (slate waste) aggregate, in Cornwall. Also, the material is used frequently in association with other walling materials. Facings to earth walls in stone, flint and brick are evident, particularly in Dorset and Hampshire.

1.13
Late Georgian cob town houses in Dawlish, Devon, built *c.* **1820**

It should be emphasised that the material included in this section on the regional distribution of earth buildings has been drawn from various published and unpublished sources, too numerous to receive individual mention, other than in certain cases. For a much fuller description of regional building types the reader is referred to Hurd and Gourley (2000) in the Bibliography.

Summary

It is clear, then, that earth buildings survive throughout many areas of mainland Britain in numbers considerably in excess of those identified in the statutory lists, and are most plentiful in Devon and nearby counties. Throughout, the material, in its various forms, is associated with small-scale buildings, but by no means exclusively so. In the south and south-west and in Buckinghamshire there are many large buildings. Some have high walls without intermediate floors, such as the Tannery at Andover, Hampshire, two chapels at Haddenham, Buckinghamshire, and various non-conformist chapels and schools in Devon (Fig. 1.15). Others, such as the rammed chalk walled villas in Winchester, Hampshire, are three or more storeys high (Fig. 1.16).

1.15
**Early nineteenth-
century non-
conformist chapel,
Cullompton,
Devon. Render
stripped, revealing
cob walls**
Photo: Peter Child

1.16
**Five-storey
rammed chalk
houses in
Winchester,
Hampshire,** *c.***1850**
Photo: Peter Walker

Earth buildings survive over a variety of geological strata, the common factor being that in all these areas, clays or chalk, together with suitable aggregates, are readily available. In many areas, among them west Wales and Lincolnshire, where suitable building stone was available only to the more affluent and timber of high quality was difficult to obtain, earth walling for some was possibly the only option. It is highly likely that before the ready availability of cheap building materials from elsewhere, and the abolition of the brick tax in 1850, earth-walled buildings were very much more abundant. The demise of earth when the alternatives became available points to it being perceived as an inferior or inconvenient material, and it is known that many earth-walled dwellings were demolished and replaced through the actions of the so-called 'improvers' of the mid-nineteenth century. However, in counties such as Devon, where alternative building materials were available, albeit not to all social classes, the relatively low drying shrinkage of the local clays coupled with the ready availability of suitable aggregates made them particularly suitable for earth walling, ensuring that their use continued until quite late in the nineteenth century.

In order to improve their durability, all traditional earth walls were provided with weather protection at head and base, usually achieved by means of a thatched roof with generously projecting eaves and an underpin course of masonry or brickwork. Later examples had slate roofs, but almost invariably the outer wall surface of these buildings had a protective coating. An external surface coating was also commonly used for higher-status domestic buildings, although many agricultural buildings, boundary walls and significant numbers of humbler dwellings remained unprotected. The original surface coating may be chalk slurry, clay render, lime/sand render, limewash, or, in some areas, coal tar and sand. A problem that has become pronounced in recent decades is the high number of structural failures due to the application of low-permeability Portland cement and plastic-based finishes, which prevent earth walls from 'breathing' and losing excess moisture through surface evaporation. This is a subject that is covered in detail in Part II, Chapter 6.

This chapter has described the main types of earth-walled buildings that can be found in Britain, and gives an account of their distribution. The following chapters examine in much greater detail the physical and behavioural characteristics of earth walls and the processes associated with their construction.

Chapter 2

The raw material: soils for use in building

Earth as a construction material

In former times, when earth building was common in many parts of Britain, earth builders, sometimes known as 'mud masons', were following centuries-old traditions. They had an intimate knowledge of the soils in their particular locality and what they were capable of. Having no knowledge of soil science or geotechnics, they relied on experience and observation. When the use of earth as a building material largely died out, at the end of the nineteenth century, much of this knowledge was lost, and is now having to be re-learned. It is considered that a basic understanding of the physical and behavioural characteristics of soil is vital, for both new earth construction and the repair and maintenance of existing earth buildings, and it is felt that this chapter forms probably the most important section of the book, for the following reasons: Firstly, for all new building projects, apart from minor domestic buildings such as summerhouses, garages, garden sheds and boundary walls, for example, Building Regulations approval will be required. This is also a requirement for major structural repairs to existing buildings. When 'conventional' materials and construction methods are being employed, compliance with the requirements set out in Approved Documents A to N and Regulation 7 can be easily achieved. The use of earth-based and some other traditional materials, as well as certain traditional building techniques, may, on the other hand, cause problems when Building Regulations approval is being sought.

Despite the obvious advantages of earth as a low-energy, sustainable form of construction, and Government advice to Local Planning

Authorities to adopt a more flexible approach to appropriate technology building methods, it is still the case that some Building Control Officers, when
faced with proposals for new earth buildings, are inclined to reject them, even
in those parts of Britain where there is a long established earth-building tradition. This is usually because the application for Building Regulations
approval contains insufficient data relating to the performance, strength and
durability of the material to enable the Building Control Officer to make a rea-
soned and informed judgement. Clearly, in the absence of any formally
approved standards and codes of practice for earth building, the officer
processing the application will need to be convinced that both the material
and the method of workmanship are capable of performing the functions for
which they are intended. It would also be helpful if the officer concerned had
sufficient knowledge of the physical and behavioural characteristics of earth
to enable he or she to interpret and evaluate the data supplied.

Secondly, it is clearly in the best interests of anyone proposing to
construct a building incorporating load-bearing earth walls to be absolutely
certain that the material they are intending to use is suitable for the purpose.
(Even a fairly modest single-storey dwelling is going to need around 125 to
150 tonnes of soil to build the external and some internal walls.)

Thirdly, and finally, in cases where serious structural problems
become apparent or are suspected in existing earth buildings, it is helpful,
when carrying out diagnostic surveys or specifying remedial works, to know
whether inherent weaknesses in the type of soil used have contributed to
the observed failure. A particular problem in all earth structures is the effect
that the presence of excess moisture can have in significantly reducing the
compressive and tensile strength of the material, an issue that is discussed
at greater length in Chapter 6.

The principal aims of this chapter are, therefore, to provide a basic
understanding of how earth works as a construction material, how different
soils can be tested and analysed in order to assess their suitability for building and how, by using geological and soils maps, and through observation, it
is possible to identify areas where soils suitable for building may be found.

Classification of soils

Depending on their intended use, soils may be classified in various ways.
Clearly the needs of the farmer or horticulturist will be quite different from
those of the civil or structural engineer. Generally speaking, the Soil Scientist
is concerned with earth as a growing medium, while the Geotechnical
Engineer is concerned with the application of the earth sciences to engineering problems, among which, at least in the context of this book, is the
use of soil as a material for building construction. In Britain, there are, at the
present time, no formally adopted standards for the analysis and testing of

soils intended for use in the construction of load-bearing walls. Most tests that are carried out are based either directly or indirectly on those contained in BS 1377. *Soils for Civil Engineering Purposes, Part 2, Classification Tests* (1990). Issues relating to the formulation and adoption of standards and codes of practice for earth construction are further discussed in Chapter 5.

The definitive soil classification maps, soil descriptions and analyses of soil types, for the whole of England and Wales, are available from the National Soil Resources Institute, at Cranfield University, Silsoe (formerly the Soil Survey of England and Wales, Rothampstead Experimental Station, Harpenden, Herts.). Similar information for parts of Scotland may be obtained from the Macaulay Land Research Institute, Aberdeen. Although the data contained in these publications are intended primarily for agricultural use, and therefore deal only with soils down to about 1.0 to 1.5 metres below the surface, they are nevertheless extremely useful in identifying the general physical characteristics of soils and assessing their suitability for building.

Undisturbed soils, that is soil samples removed from the field by means of core drilling or the excavation of trial pits, may, following field or laboratory testing and examination, be classified in various ways, according to their intended use. If the material is to be used for building one needs to be able to assess its suitability in terms of its potential performance and durability, and in order to do this some or all of the following information will be required:

1   **Soil type.** This can be determined initially by reference to the Soil Survey maps referred to above, and would include details of the soil's parent material (underlying geology) and mineralogy, its stratigraphy (thickness of, and variations in, soil horizons) and its colour, which may be important in cases where the wall faces are to remain exposed or when it is intended to apply mud plasters to internal or external walls.

2   **Particle size distribution (psd).** It is essential to know the composition of any soil intended for use in building. Under the system adopted in BS 1377, Part 2, soil particles are classified according to their diameter, from the largest down to the finest. Stones and gravels are particles greater than 2.0mm, sands include all particles between 2.0mm and 0.063mm (63 microns, symbol μm) and any particle below 63 microns diameter is classed as 'fines' – silt and clay (see Fig. 2.1). All the above, apart from the clay fraction, are further subdivided into coarse, medium and fine. A typical particle size distribution chart is shown in Fig. 2.2.

3   **Percentage clay fraction.** Clays are the very finest particles, with a diameter <2μm, according to BS 1377. In fact, some coarser clay particles may be seen, on scanning electron microscopy

photographs, to have a diameter between 5 and 10µm, and in the USA the upper limit is taken as 5.0µm (USDA classification). Clays are of crucial importance in determining the performance characteristics and behaviour of soils, as they act as a binder or matrix cementing the coarser aggregates together in a cohesive and homogeneous mass.

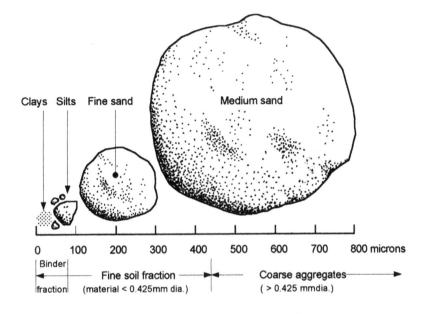

2.1
Clays and aggregates – relative particle size diameter

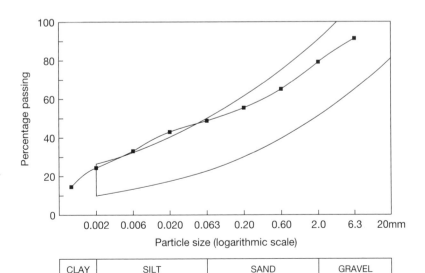

2.2
Particle size distribution (psd) chart showing, within the curved lines, soils suitable for cob or mudwall construction.
The psd curve plotted is that of a clayey sand soil used for a test programme carried out at Plymouth University

CLAY	SILT			SAND			GRAVEL	
	Fine	Medium	Coarse	Fine	Medium	Coarse	Fine	Medium

4 **Soil consistency.** This refers to the plastic properties of clay-based and chalk soils, which enable them, when mixed with water, to be moulded, sculpted and formed into load-bearing walls or used as plasters and external renders. Soils that are either non-plastic or of very low plasticity are, as a general rule, insufficiently cohesive to be used for building.

5 **Soil expansiveness.** This refers to the swelling and shrinkage of soils during the wetting, mixing and drying-out process, and is largely dependent on the amount and type of clay present. Soils containing a high proportion of clay are unsuitable for mass earth walling because of their high volume shrinkage, but may be suitable for the fabrication of unbaked earth blocks or bricks.

6 **Compressive strength.** Some indication of the load-bearing capacity of a mass earth wall will be a requirement when applications for Building Regulations approval are submitted, for obvious reasons. Testing the unconfined compressive strength of a soil sample, usually a 100 or 150mm diameter cylinder with a length twice that of its diameter, can only be carried out in a suitably equipped soils laboratory. Specimens are usually tested at their equilibrium (air-dry) moisture content. However, in cases where it is considered necessary to establish the critical moisture content, or minimum wet strength for a given load, of a soil mix, then specimens at various moisture contents would need to be tested.

7 **Soil density.** It is useful to know the air-dry density of a soil for three reasons: Firstly, it is needed in order to calculate foundation and ground loadings; secondly, for estimating the amount of material required to construct the building, especially in cases where soil is to be transported from an outside source; and, thirdly, in order to assist in calculating the thermal conductivity of the walls (a specific requirement for new-build and conversion projects). In addition, knowing the density of a soil enables one to estimate its porosity – volume of voids expressed as a percentage of total volume.

Soil classification in detail: identification, analysis and testing

Identification of soil type

It should be stressed that topsoil, the topmost part of the soil down to a depth of 400 to 450mm, should not be used for building because it normally contains far too much organic material. Using potentially productive topsoil for building would, in any event, represent waste of a valuable resource. For those aspiring to build with earth, a lot can be learned, through practical experience and by observation, about the soils in one's local area and one does not need to be a soil scientist or geotechnical engineer to realise that soil characteristics, even within a small geographical area, can show enormous variation. Taking two examples at extreme ends of the spectrum, one might compare a 'light' soil, comprised almost entirely of fine to medium sand, with a 'heavy', silty clay containing numerous flint and gravel particles. Experience would show that the sandy soil would be easy to cultivate and very free-draining, drying out rapidly in warm, windy conditions, while the clay soil would be heavy, sticky and glutinous, and in winter especially, impossible to cultivate. Moreover, it would retain moisture, even in summer, because drainage would be impeded by the low moisture permeability of the clay. In dry conditions the sandy soil would be easy to break down and cultivate, while the clay soil could only be broken up with a pickaxe or mattock. Although the physical characteristics of the two soils are totally different, both would be equally unsuitable for mass earth construction without modification, though for different reasons, all of which will be explained and clarified below.

Around 15 per cent of the land area of England has chalk at or near the surface. Much of this is classified geologically as Upper Chalk and forms part of the Upper Cretaceous formation. Although one might describe this material as subsoil, because of its proximity to the surface, it is, strictly speaking, a soft rock, and has physical and behavioural characteristics rather different from those of clay-bearing subsoils. Chalk has been used for building construction in the southern counties of England for at least two or three hundred years. Because the Upper Chalk is comprised almost entirely of fine calcium carbonate particles, less than 0.1mm diameter, and contains virtually no clay, test procedures 2, 3 and 5 outlined above are not applicable. Chalk is discussed in greater detail at the end of this chapter.

It should perhaps be noted at this stage that there are some practical guides and handbooks on earth building which contain details of simple field tests that can be carried out in cases where soils laboratory facilities are either unavailable or unaffordable, for example in developing or 'third world' countries (Norton, 1997). However, some of the tests described, especially if carried out by people with little experience of earth building, may be of only limited use in the sense that the data they provide may not be of sufficient accuracy or credibility to convince a sceptical Building Control Officer. On the other hand, such tests can prove very useful when 'prospecting' for suitable soils, enabling one, at an early stage, to eliminate those soils that are clearly unfit for use.

Particle size distribution (psd)

Soils considered most suitable for mass earth walling are those that are termed well graded. This is an engineering term, which refers to soils comprised of particles of many different sizes, having a uniform distribution from coarse to fine. In other words, no single particle size is predominant. Particle size distribution characteristics of soils can only be accurately determined by means of dry and wet sieving, and sedimentation analysis.

The preferred size of a soil sample intended for psd analysis is dependent on its stone and gravel content. For soils that contain a large proportion of coarse aggregates (material > 2.0mm) a sample weight of 600g is recommended, while for soils comprised mainly of fine material 300g is considered sufficient. The sample is first broken down, though not necessarily pulverised, in order to separate out the coarser aggregates. An ordinary garden or mason's sieve with a 5 to 6mm mesh will retain the coarse gravel and stone fraction, which can then be weighed and set aside. The remaining soil is then immersed in water, to which a heaped teaspoonful of sodium (Na) is added, either sodium chloride (common salt) or, in a laboratory situation, Calgon (sodium hexametaphosphate). The effect of adding sodium ions to the water is to disperse flocculated clays (this process is illustrated in Fig. 2.9 and further explained in the section on clay below). After soaking, and stirring occasionally, for one hour the soil is poured or washed through a nest of sieves into a container, in which will be contained the finest, silt and clay, particles.

If the procedures described in BS 1377 are used as a general guide, 200mm diameter sieves with the mesh sizes shown in Table 2.1 should be used. The material retained on each sieve is oven-dried and weighed, so that it can be represented as a percentage (by weight) of the whole sample. As noted above, the material remaining, which has passed through the 63µm sieve, is silt and clay. If this material, which will be immersed in water at this stage, is allowed 24 hours to settle, the water can then be carefully poured or siphoned off and the sediment oven-dried. It can then be crushed, using a pestle and mortar, and the resulting powder used to carry out a sedimentation analysis, should this be required. Figure 2.2 shows the psd curve for the soil type also featured in Figs 2.3 and 2.5. The upper and lower lines on the chart indicate the range of soils considered suitable for mass earth construction. Any soil whose psd curve falls within this area would be suitable for cob construction. For rammed earth construction, however, the clay content would need to be lower, 5 to 10 per cent of material < 0.002mm (2µm).

Table 2.1 Wet sieving

Mesh size	Material retained
2.00mm	Fine gravel
0.60mm	Coarse sand
0.20mm	Medium sand
0.06mm	Fine sand

Sedimentation analysis

Some earth-building manuals describe a simple field test known as the 'jar' test, the purpose of which is to identify the proportions of sand, silt and clay present in a soil sample. The procedure, simply described, involves filling a screw-top glass jar or similar container with one-third fine soil (< 2.0mm) and two-thirds water, adding a little salt then shaking the jar vigorously, allowing the soil to disaggregate, shaking again, and then observing and measuring (a) the time it takes for the particles to settle and (b) when settlement is complete, the relative proportions of sand, silt and clay particles. The test is too crude to yield accurate results, but may be useful in identifying general soil characteristics. For example, if all the soil has settled to the bottom of the jar in less than 30 minutes, leaving the water above clear, then it contains no clay (this would be quite unusual). However, this test does introduce an important concept, that of settlement velocity. This is based on Stokes' law, which enables one to predict the speed at which particles of a given diameter and specific gravity (particle density), at a given temperature, will fall through a liquid. In the jar test referred to above, all the sand particles, other than the very finest, will have settled to the bottom of the jar within around 30 seconds. The mean settlement velocity of silt is 10mm per minute and that of clay 0.2mm per minute (Einteche, 1964). Knowing the settlement velocities of particles of different diameters has enabled the development of laboratory tests, using specialised equipment, which are capable of grading silts into fine, medium and coarse particles, and accurately measuring the all-important clay fraction.

The clay fraction

There are some who would claim that knowing the precise amount of clay present in a soil is not absolutely vital, and that plasticity characteristics derived from the soil consistency tests, described below, are all that is required. However, clay is as important in mass earth construction as Portland cement is in reinforced concrete structures. Although a degree of cohesion can be achieved through compaction, which forces the soil particles into close association so that they cohere by means of inter-granular friction, or interlocking, this in itself cannot provide the bonding strength required to support the imposed loads of a two- or three-storey building, or even a two-metre-high boundary wall. Moreover, in practical terms, the ratio of silt to clay may have an important influence on the soil's performance, as experience has shown that earth walls built from silt-rich soils are particularly prone to erosion and weathering. It is for these reasons that any standards or codes of practice for new earth building that might be adopted in the future are almost certain to include recommendations relating to minimum clay content.

Soil consistency

Cohesion and plasticity are two of the most important properties of soils intended for use in building. A concise and lucid explanation of the two terms, which are often thought of as being synonymous, is given in Carter and Bentley (1991) where the authors conclude that, 'Whereas plasticity is the property that allows deformation without cracking [rather like putty or Plasticine], cohesion is the possession of shear strength that allows the soil to maintain its shape under load, even when unconfined'. Both plasticity and cohesion in soils are determined by (a) their particle size distribution characteristics, (b) the size of the clay fraction, (c) the state of packing of the soil particles, which is dependent on the extent to which the soil mass has been compacted, and (d) the amount of free water present in the soil. In general terms, a non-plastic soil, one comprised almost entirely of sand and silt for example, will also be non-cohesive and therefore quite unsuitable for building without modification. Furthermore, walls constructed from soils of low plasticity tend to have little resistance to the effects of excess moisture and will, therefore, be more susceptible to structural failure. It is for these reasons that the tests described here, known as the Atterberg limit tests, are considered so important.

Figure 2.3 shows how moisture content relates to volume change and plasticity in a clayey sand soil overlying the Permian breccia rock formation of south Devon (Greer, 1996). The soil is by no means typical, but represents a material ideally suited for cob building, though not for rammed

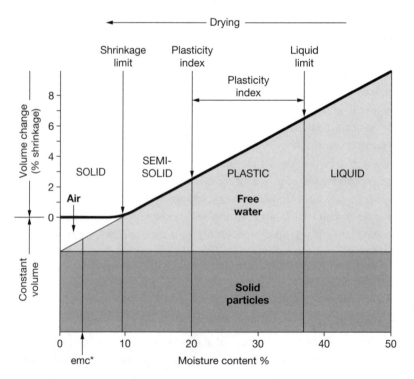

2.3
Volume change and consistency related to moisture content in soils
*emc = equilibrium (air-dry) moisture content (based on Greer, 1996)

earth as it contains too much clay, at 24 per cent by weight. Reading from left to right, the soil is first shown in an oven-dry state, with all free water driven off from the pore spaces (soil voids), which are filled with air. At equilibrium moisture content (emc) the soil is air-dry, and its moisture content, at about 3 to 4 per cent, is in equilibrium with that of the surrounding atmosphere. Above the shrinkage limit (moisture content 9.5 to 10 per cent) continued ingress of free water will lead to a gradual increase in volume, but no significant loss of strength or cohesion. As the plastic limit is approached there is an increasing loss of cohesion, and therefore compressive strength, because the electro-static and hydrogen bonds between the clay particles are being broken down – forced apart by excess free water entering the soil voids, allowing the soil to deform under load. At the plastic limit (20 per cent moisture content) the soil will begin to deform progressively and irreversibly, and, as the moisture content increases, will develop vertical or diagonal shear cracks, at which point structural failure will occur. In load-bearing walls, the liquid limit, the point at which a soil will begin to flow under its own weight, may appear to be of little more than academic interest. However, knowing a soil's plasticity index (PI) does enable one to predict the likelihood of either sudden shear failure through rapid crack propagation, resulting from a modest increase in moisture content, or slower, more progressive failure as a result of plastic deformation.

It has been estimated that all soils, at the plastic limit, have a shear strength of at least $100kN/m^2$ but only $1kN/m^2$ at the liquid limit (Carter and Bentley, 1991). Clearly therefore, it is the plasticity index that will determine the behaviour of a damp soil, and the lower the PI (chalk, for example, has a PI of about 5) the greater is the risk of sudden structural failure. It should be noted that, in order to comply with the standards set out in BS 1377, only the fine soil fraction, particle size < 425µm, is used for soil consistency tests. Some simple field tests for assessing, but not accurately determining, the plasticity of soils intended for building are described in Norton (1997).

Soil expansiveness

Knowing the extent to which a soil will swell and then shrink during the mixing, building and drying-out process is crucially important, and may either determine the most appropriate construction method to be adopted or indicate the need to modify a soil in order to reduce shrinkage where the material is to be used for mass walling. Generally speaking, the higher the clay fraction, the greater will be the soil's expansiveness. Depending on their mineralogy, some clays are more expansive than others. This topic is further discussed below. The change in volume that occurs in soils as a result of increasing moisture content was discussed in the previous section and is illustrated in Fig. 2.3. Soil expansiveness may be measured fairly simply by means of a shrinkage box. This can be constructed from 15 or 16mm planed timber, or melamine-faced plywood, and is essentially a long, open-topped box of internal dimensions 600 × 50 × 50mm. The soil, with fibre added if required, is mixed with water to the consistency at which it would be placed on the wall. It is then pressed or rammed into the box and left for several days to dry out. The gap between the air-dried soil sample and the end of the box can then be measured and the linear shrinkage expressed as a percentage of 600mm.

Compressive strength

The maximum compressive strength of a soil is the point at which shear failure occurs under an applied load – sometimes known as failure stress – and is normally measured in either kilonewtons per square metre (kN/m^2) or newtons per square millimetre (N/mm^2); $1,000kN/m^2$ is equal to $1N/mm^2$. In some European countries compressive strength is measured in kg/cm^2 ($20kg/cm^2 = 2N/mm^2$). In this book compressive strength is given in kN/m^2. Measuring the unconfined compressive strength of air-dry specimens of reconstituted soil, in the form of cylinders either 100 × 200mm or 150 × 300mm (see Fig. 2.4) provides, at best, an approximation of the load-bearing capacity of a mass earth wall.

However, testing full or even half-scale walls to destruction is hardly a practical proposition. Some soil samples have been tested using a procedure similar to that normally employed for testing concrete cubes, which gives a height-to-width ratio of 1:1 for the test sample as opposed to the 2:1 ratio normally adopted in soil testing, which tends to give a higher strength at failure and may therefore be misleading. In these cases, a correction factor would need to be applied in order to provide comparative results. Laboratory tests carried out in Australia on stabilised rammed earth samples (Middleton and Schneider, 1992) and at Plymouth University, on cob samples (Saxton, pers. com.), would suggest that test cubes show a compressive strength at failure between 20 and 25 per cent higher than that of cylinders with an aspect ratio of 2:1. The correction factor proposed by Middleton and Schneider in this case is 0.78. The overall compressive strength of an adobe or clay-lump wall is more difficult to calculate because, unlike cob or rammed earth, it is not a monolithic form of construction. For example, tests carried out at Plymouth University, on cob blocks containing 8 per cent lime putty, showed a compressive strength of $2,500kN/m^2$ for a single block laid flat but only $540kN/m^2$ for a stack of three blocks laid in a mud mortar. Comparing mass, monolithic earth walls with those of small unit construction – adobe and clay lump for example – it is clear that, despite the strong adhesion of earth-based mortars, the joints between individual bricks or blocks represent points of potential weakness. Research carried out in Peru to compare the

earthquake resistance of adobe and rammed earth walls showed that, in diagonal compression tests designed to simulate seismic activity, adobe walls showed 40 per cent less resistance to shear failure than those constructed of rammed earth (Vargas Neumann, 1993). Accepting a standard form of laboratory test procedure which can easily be replicated and provide comparative data would enable the adoption of national standards relating to minimum dry and wet strengths in soils intended for use in building.

Dry compressive strength can vary considerably, according to the grading of soil particles, the size and mineralogy of the clay fraction, and to the degree of soil compaction achieved. In cob walls, strength will be in the range of 600 to 1,100kN/m^2, but can be higher – up to 1,400kN/m^2 in clay-rich soils, for example. In rammed earth walls, 800 to 2,000kN/m^2 is considered normal, though strength can be higher in cases where steel-reinforced shuttering and pneumatic rammers are employed. The high density and compressive strength achieved in rammed earth enables quite slender walls to be constructed, 300 to 450mm wide, thus using far less soil than would be required for a cob wall of equivalent height. Compressed soil blocks made using a hydraulically assisted block press can also achieve high compressive strengths, around 1,500 to 2,000kN/m^2. Compressive stress at plinth level in a traditional two-storey cob house with 550 to 600mm thick walls, is in the region of 80 to 100kN/m^2. A series of tests were carried out at Plymouth University to examine the relationship between compressive strength and

2.4
Machine for testing the unconfined compressive strength of soils
Photo: ELE International

2.5
Relationship of UCS (unconfined compressive strength) to moisture content in a fibre-reinforced, clayey sand soil

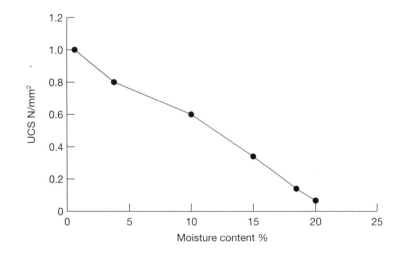

moisture content in a fibre-reinforced soil with similar psd characteristics to that shown in Fig. 2.2 (Saxton, 1995). The test results showed, as expected, that strength in cob walls will decrease with increasing moisture content in a more or less linear fashion (see Fig. 2.5) down to around the plastic limit, and that a further increase in moisture content beyond this point will result in failure as a result of plastic flow. Figure 2.6 shows two cob test cylinders after failure. The damp sample on the left (moisture content 10.5 per cent dry weight) has 'barrelled' and failed as a result of plastic deformation. More

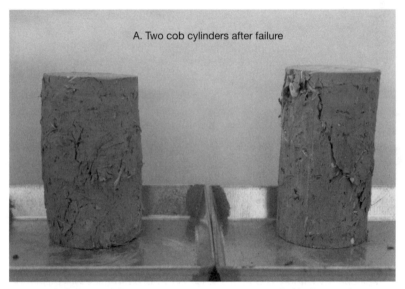

A. Two cob cylinders after failure

1. Damp sample   2. Air-dry sample

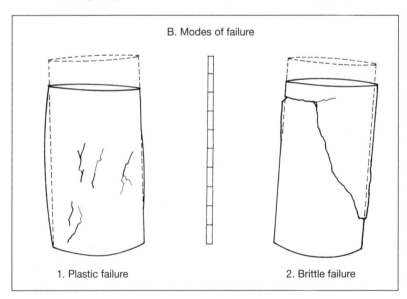

B. Modes of failure

1. Plastic failure   2. Brittle failure

2.6
Typical unconfined compressive test specimens
Vertical scale = 200mm

or less vertical cracks have developed as a result of tensile stress under load, allowing the soil mass to separate into discrete pieces, whereas the air-dry sample, on the right, has suffered a type of 'brittle' failure through the development of a diagonal shear crack, typically forming an angle of approximately 60 degrees.

Soil density and porosity

Density is simply weight divided by volume, and is usually expressed in either kilograms per cubic metre (kg/m^3) or grams per cubic centimetre (g/cm^3) – 1Mg (1,000kg)/m^3 = 1.0g/cm^3. To avoid confusion, kg/m^3 has mostly been used in the text and all densities given refer to air-dry bulk density at equilibrium moisture content, except where stated. Density in reconstituted soils varies according to particle type and size distribution, and to degree of compaction. A soil comprised mainly of clay, silt and fine sand, for example, will have a low density of around 1,600kg/m^3, whereas a soil containing a large proportion of stone, gravel and coarse sand would have a much higher density – 1,800 to 2,000kg/m^3. Rammed earth, because it is subjected to heavy compaction, usually has a density of around 1,900 to 2,200kg/m^3, and chalk, depending on moisture content at mixing and degree of compaction, will vary from 1,450 to 1,550kg/m^3. Porosity can be defined as the percentage pore space within a given volume of soil; known also as total or absolute porosity, which is a measure of the total volume of voids from which water can be removed. Dry density refers to oven-dried soil, and particle density to the specific gravity of the solid particles, which, in soils, is generally accepted as being around 2.65 to 2.7 (2,650 to 2,700kg/m^3). Thus, it may be seen that both density and porosity in soils are determined by particle size and packing, and by degree of compaction. Knowing a soil's dry density enables one to estimate its porosity and to predict its thermal performance. Particle size and porosity are also important in determining the permeability of soils, a topic which is discussed in greater detail in Chapter 6.

Clays and aggregates, including chalk

Stone and gravel

The stone and gravel fraction includes all material greater than 2.0mm diameter, and can vary from low density, porous, soft shale, oolitic limestone and chalk, up to high density, largely impermeable, sandstones, quartz, flint, and the older (Devonian-age) limestones. Excessive amounts (over about 45 per cent) of coarse gravel and stone in soils can cause problems for three reasons. Firstly, and most importantly, in soils containing large quantities of stone and gravel, inter-granular contacts between the all-important clay particles are much reduced. This results in lower cohesion, which leads, in turn, to a reduction in compressive strength. Tests carried out on adobes (compressed soil blocks containing 3 per cent bitumen) in the south-west USA showed that blocks containing coarse aggregates had a compressive strength of 280psi (2,000kN/m^2), while those containing only fine aggregates, < 10mm diameter, reached 400psi (2,750kN/m^2) before failing, representing an overall gain in strength of 30 per cent (Tibbets, 1992). Secondly, large stones can cause problems in mass earth walls, especially when they appear on the finished face of a wall that is to remain unrendered. In cob walls, stones can be hooked out during the paring down process, leaving unsightly cavities, which may attract the unwelcome attention of masonry bees and, more seriously, encourage rainwater to enter the wall face, causing rapid localised erosion. To a certain extent, the same is also true of rammed earth walls constructed of silt-rich soils, particularly in cases where they are exposed to driving rain. Thirdly, experience has shown that cob blocks containing large stones, over about 25mm in diameter, have a tendency to break up when being transported and handled on site.

Sand

Sand-sized particles are non-cohesive and, apart from those sands that contain fragmented seashells and other calcareous materials, mostly non-porous. Sand particles may be either angular or rounded, according to the extent to which they have been abraded as a result of transportation or weathering. Soils containing angular sand grains will tend to be less porous than those containing rounded grains, especially when compacted. This is because angular grains, when under pressure with free (pore) water to act as a lubricating layer, are forced into close association with adjoining grains, and, to a certain extent, interlock; thus achieving a degree of cohesion through what is known as internal friction, a phenomenon which goes some way towards explaining the high compressive strengths achieved in rammed earth walls, in which the soils are heavily compacted. If the sand grains in a soil are angular rather than rounded, and are also well graded, the fine sands, silts and clays (particle diameter < 425μm and known as the fine soil fraction) provide a cohesive matrix within which the coarser particles are contained.

Silt

Silts are comprised mainly of fine quartz material, and are often rounded rather than angular in form. Although not cohesive in the same way as clay minerals, it is thought that a limited degree of cohesion in silts may be achieved through surface friction and inter-particle moisture films (even in 'dry' walls, there is always a certain amount of free water present in the micro-pores – voids < 50μm diameter). In soils used for building, a ratio of

1:1 or 1:2 clay to silt, would be considered optimal. However, many soils, described as 'silt-rich' in the text, have clay/silt ratios much higher than this – up to 1:4. Experience has shown that walls built from such soils, where they are unrendered and occupy exposed positions, are very prone to surface erosion (see Fig. 2.7). This is because the largely cohesionless fine silt particles can, in the absence of a significant clay fraction, easily be washed away. This view has been confirmed by the results of accelerated weathering tests recently carried out in Germany and Australia (Heathcote, 1995; Minke, 2000). It should be noted that some of the finest particles classed as silt, those in the range from 2 to 10µm, may, in fact, be clay minerals. This issue is further discussed below.

2.7

Partly collapsed cob wall, showing the effects of long-term weathering on a silt-rich soil

Clay

It might be supposed that all one needs to know about clay is that it is cohesive, and acts as a kind of cement, binding the coarser soil particles together. However, it is considered that a simple, but hopefully clear, explanation of how clays work is essential if one is to understand the nature of soil as a building material, and why clays play such a crucial role in soil cohesion.

Clay particles are so small that they can only be properly examined by means of an electron microscope, Their internal molecular structure is made up mainly of layers of silica, alumina and, to a lesser extent, magnesium, with oxygen and hydroxyl ions providing inter-layer bonding. Most clay particles are plate-like in form and more or less hexagonal in shape (see photographs in Fig. 2.8). With a thickness one-tenth to one-twentieth of their diameter, they have a surface area many thousands of times greater than that of even the finest sand grain. Cohesion between individual particles is achieved by means of strong ionic bonds and by hydrogen bonds, which, although they are relatively weaker, form an integral part of the clay's structure. Reference to the much simplified diagram in Fig. 2.9 will show that the surface of each clay particle has a net negative charge, while its edges are positively charged, and that each particle is surrounded by an adsorption layer, or film, of viscous 'bound' water, which has a net positive charge. The free water surrounding the clay particles contains both anions and cations, but it is the positively charged cations that are attracted to the surfaces of the particles, forming the hydrogen bonds referred to above. Also shown in Fig. 2.9 is the effect that excess free water has on the disposition of clay particles. As more free water enters the soil voids, the relatively weak hydrogen bonds between the clay particles are broken down, forcing them apart and, ultimately, into suspension, at which point the soil will behave as a liquid slurry. This condition is shown on the left of Fig. 2.10, where the clay particles are in a dispersed state. They have assumed a layered pattern and a net repulsive force exists between them. When the excess free water is removed, as shown on the right of Fig. 2.10, individual particles are able to move towards each other and join edge to face, positive to negative. The hydrogen bonds are re-established, and a condition known as flocculation exists. It should be noted that in an oven-dry soil, where all free water has been driven off, cohesion is maintained by the bound water layers alone. Adsorbed water can be removed from the soil only at temperatures exceeding 300°C.

Not all clays behave in the same way, however, and much can depend on the chemistry of cations contained in the pore water, as well as on its pH (acidity or alkalinity). Sodium ions can act as a dispersant, for example, and calcium will encourage flocculation. The three clay types most commonly found in northern European soils are, in order of importance, the clay mica group, generally known as illites, the smectite group, which includes montmorillonite, and kaolinite, perhaps better known as china clay. Kaolinite has a relatively simple molecular structure and is the most stable of the clay minerals, showing only minor swelling and shrinkage (expansiveness). Clays of the mica group, which comprise a large part of the clay-sized fraction in most common British soils, are generally more complex and variable in nature than kaolinite, and less well understood. Their inter-layer molecular bonding is weaker than that of kaolinite and they are moderately

(a)

(b)

(c)

2.8
Micro-particles: scanning electron microscope photographs of clay and chalk particles. Above, left, flakes of clay mica (illite) and on the right a stacked arrangement of kaolinite – the bars represent five microns. Below, left, are shown the minute marine organisms called coccoliths, of which chalk is mainly formed. In this case the bar represents 10 microns
Photos from Rowell, 1994

expansive. The smectite group of clays is characterised by very weak inter-layer bonding, which means that these clays will readily absorb water and are, therefore, very expansive. Clay mineralogy is an extremely complex subject, with which the reader need not be overly concerned because

2.9
Simplified diagram showing electro-static bonding between clay particles

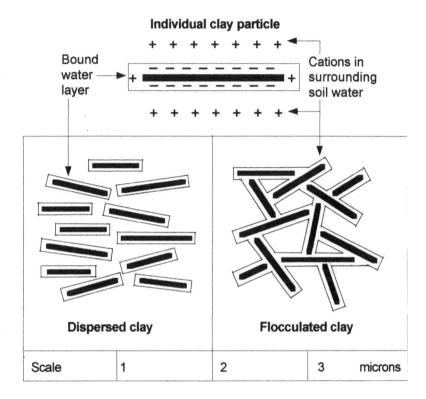

Notes: 1. Clay in supension (moisture content above liquid limit). All voids (pores) filled with water. 2. Pore water evaporating. Moisture content above plastic limit. Soil can be moulded and re-formed. 3. Oven dried. All pore water removed and voids now filled with air

2.10
Diagram showing the effects of moisture on the behaviour of clay particles
Adapted from Artifacts: An Introduction to Early Materials and Technology by Henry Hodges, London: Gerald Duckworth & Co. Ltd, 1989, p. 22, Fig. 1 (1, 2 and 3), with kind permission

swelling and shrinkage in soils, although important in the building process, can, as noted above, be fairly easily measured and, if necessary, controlled by modifying the soil.

Chalk

Chalk occurs widely in the south of England, but has been used for building construction mainly in the counties of Dorset, Hampshire and Wiltshire, the area known as Wessex. In eastern England the chalk is mostly covered by later, mainly glacial, 'drift' deposits, though the underlying chalk has had a significant influence on the structure and physical characteristics of subsoils in East Anglia, Lincolnshire and east Yorkshire. The chalk was laid down during the Cretaceous period, between 60 and 140 million years ago, when most of the land area of present-day Britain lay beneath a warm, tropical sea. In places, the chalk beds are up 500 metres in depth. The Upper Chalk, which is the most widespread geographically and is therefore the material that was most usually employed for building construction, is, in effect, a soft limestone rock of exceptional purity. Because it is comprised almost entirely of micro-scopic marine organisms, none of which are larger than 0.1mm in diameter (see photograph Fig. 2.8), its calcium carbonate content is very high, in the region of 96 to 98 per cent. With increasing depth the chalk becomes less pure and in the Middle, and, in particular, the Lower Chalk formations, increasing amounts of silica and clay minerals are found; the material has a higher density than the Upper Chalk, is often harder and therefore more difficult to extract. These are known geologically as the chalk marls or grey chalks and, because they are found near the surface to a much lesser extent than the Upper Chalk, have probably not been used extensively for building, other than in the form of cut (quarried) blocks, such as the 'clunch' from the Lower Chalk beds of Cambridgeshire. In the chalk downland areas of southern England, chalk is often found within ploughing depth, 300 to 400mm from the surface and is usually in fragmented form, having been broken down by the action of rain and frost as well as by cultivation. Chalk differs from normal clay-based subsoil in several important respects. Firstly, because calcium carbonate particles, even the smallest clay-sized fraction, are electrically neutral – having neither a negative nor a positive charge – no ionic or hydrogen bonding can take place between them. There is no bound water layer surrounding individual particles, so that the voids contain either air or free water, usually a mixture of both unless the material is saturated. Chalk, therefore, would appear, in an electro-chemical sense, to be non-cohesive. Secondly, the almost complete absence of clay minerals means that the material is much less expansive than most clay-based soils, especially when it is rammed into shuttering at a moisture content of around 22 to 23 per cent. Thirdly, because chalk is comprised entirely of very fine particles, it is micro-porous, and therefore has a high porosity; 30 to 50 per cent of its volume is made up of air voids, so it has a high water storage capacity and, when drying out, loses water very slowly. Much research has been carried out into the behaviour of *in situ*, undisturbed chalk in connection with major civil engineering projects such as the Channel Tunnel. However, much less is known about how the material behaves when it is reconstituted and used to build freestanding structures. Clearly, the survival of numerous buildings constructed wholly or partly of chalk in southern England provides evidence of the durability of the material and its potential for new construction.

So, how is cohesion achieved in what appears to be a largely non-cohesive material? The fact that chalk behaves plastically, albeit over a very small range of moisture content (typical plasticity index of five) must indicate that a degree of cohesion exists, and that 'densification' (close packing of particles as a result of compaction) and the presence of pore water both play a crucial role in this. It has been suggested that 'true' cohesion can only be achieved in Upper Chalk by the addition of a clay binder or other cementing agent (Northmore, British Geological Survey, pers. com.), and this view would seem to be supported by the fact that many of the 'chalk cob' buildings in Wessex contain an admixture of clay subsoil, usually in the proportion of 1:2 to 1:3 soil to chalk, together with organic fibre reinforcement. The other main-stream technique, also found in Wessex, mainly in the Winchester and Andover areas, is an adaptation of the *pisé de terre* method, in which damp, rather than wet, fragmented chalk is rammed into timber shuttering to form fairly slender monolithic walls (Pearson, 1992). According to Pearson, chalk is well suited to this form of construction, in which dense packing of the particles is achieved thus increasing its compressive strength. However, it would appear that this densification process needs to be supplemented by the presence of moisture in the finer chalk particles, those less than 0.4mm diameter, enabling them to act as a matrix within which the coarser particles are contained. Experience gained in the use of reconstituted chalk in civil engineering works – embankments and road construction – would suggest that when the moisture content of the fine chalk matrix exceeds 20 per cent a rapid loss of strength will take place (Northmore, op. cit.). Tests carried out on samples of reconstituted, fibre-reinforced Upper Chalk from Blandford in Dorset showed a typical plastic limit of 25 per cent, a liquid limit of 30 per cent, an air-dry compressive strength of 800kN/m^2 and a dry density of 1,490kg/m^3. By adding one part of clayey sand soil to two parts of Upper Chalk plus chopped straw, a compressive strength of 1,110kN/m^2 has been achieved at a dry density of 1,540kg/m^3, an increase in strength of around 38 per cent over the pure chalk cob specimen. Samples of Upper Chalk from Dorset and Berkshire show the compressive strength of fibreless rammed samples to vary from 500kN/m^2 (moderate, manual compaction) up to around

1,100kN/m^2 (simulated pneumatic compaction). Air-dry densities are never higher than about 1,550kg/m^3. Experience would suggest that the strength of reconstituted chalk may increase over time, provided it remains damp, for reasons that are not yet clearly understood, and that resistance to erosion and weathering in well-compacted chalk/straw blocks is as good, if not better, than that of clayey sand cob blocks. Unfortunately, no field or laboratory testing of material from standing buildings appears to have been carried out. Curiously, given the high porosity of chalk, the average air-dry moisture content of chalk cob walls, measured in the field, has been found to be around 2 per cent, almost half that of clay soil cob walls (Trotman, 1995).

The use of organic fibrous reinforcement in mass earth walling and the modification of soils in order to improve their performance are both dealt with in the following chapter.

Chapter 3

Earth construction 1: preparation and mixing

Constructing mass earth and earth block walls

It is not the intention, in this and the following chapter, to deal with every aspect of earth building in great detail, nor to provide a comprehensive step-by-step building manual, but to describe and illustrate general principles, and to focus on earth wall construction in the context of contemporary building practice and the need to comply generally with Building Regulations. Only in cases where certain building elements need to be modified or adapted to suit earth wall construction will they be mentioned.

First, find your soil . . .

Every year in Britain millions of tonnes of earth, much of which could be used for building, is removed from construction sites and transported, at great expense, to inert waste tips, where it is simply dumped or used as landfill. For the prospective earth builder, therefore, finding suitable raw material should not pose a problem. Of course, ideally one would employ subsoil dug from the site upon which the new building is to be located. However, this is usually only possible on a site that is either large enough to allow the ground to be remodelled or which occupies sloping ground containing deep subsoils. If the *in situ* soils are entirely unsuitable for building, one has two options; either the soil can be modified or material can be obtained from elsewhere. If, on the other hand, the soils are suitable but are too shallow to provide sufficient material, they will need to be supplemented by similar soils from nearby. It is usually possible to find a source of suitable soil within five miles (eight kilometres) of the site. It is sometimes the case that subsoils have been formed in such a way that they are comprised of shallow layers, or horizons, of materials that show quite different characteristics, so that an upper horizon – immediately below the organic topsoil – of clayey or silty sand for example, may be underlain by shale, gravel or a layer of stony, impermeable clay. The Soil Survey maps and publications referred to in Chapter 1 will provide general guidance concerning soil characteristics down to about 1.0 to 1.2 metres below the surface in a particular location. However, it will always be necessary to either dig trial pits or take core samples from any proposed building site as part of the preliminary design and specification process, and it is at this stage that the suitability of the subsoil for building can be fully assessed. Where there are two or more layers of soil, the physical characteristics of which are quite different, samples from each layer will need to be tested in order to determine whether, when mixed, they will form a composite material suitable for building. As already noted in Chapter 2, a soil may be appropriate for one construction method but not for another, in which case one can either adopt the building method for which the soil is best suited or modify the soil to suit the preferred form of construction. Soils may be modified in various ways according to their physical characteristics and their intended use, the need for modification having first been established by assessment of data contained in the results of the test procedures described in Chapter 2.

Soil stabilisation

A great deal of the internationally published literature on the subject of earth building is based on the use of what are known as 'stabilised' soils, by which is meant soils whose strength, durability and resistance to moisture have been improved by the addition of Portland cement, lime or bitumen. In certain areas of the world where, for example, seasonal monsoon conditions may cause severe erosion and serious flooding, and in regions where seismic activity is an ever-present threat, the use of stabilised soil in building construction is probably justified. However, in temperate, less extreme climates it should be possible, by exploiting fully the inherent physical properties of soils and by careful blending and mixing where necessary, to construct strong, durable buildings without recourse to stabilisation. Adding up to 6 per cent, by weight, of Portland cement to a sandy soil will result in an (often unnecessary) increase in compressive strength, and a high resistance to moisture penetration and surface erosion. However, during this process the soil undergoes a fundamental and irreversible chemical change so that it is no longer recyclable, becoming, in effect, a sort of 'brown concrete'. In order to stabilise clay-rich soils, which are, in themselves, usually sufficiently cohesive, 3 to 10 per cent by weight of non-hydraulic lime may be added. The effect of this is to reduce plasticity and shrinkage, and improve resistance to moisture, thereby increasing wet strength.

Although a chemical reaction takes place between the clay fraction and the calcium hydroxide, of which the lime is largely comprised, this is not as drastic as the effect of cement, and may, in the long term, be reversible. Lime is mentioned from time to time throughout the book, particularly in connection with plasters and renders for earth walls, so it might be useful, at this stage, to describe briefly the three types of lime in current use. Non-hydraulic, high calcium lime comes in two forms, either as lime 'putty' or in bagged, powdered form, when it is known as hydrated lime. High calcium limes will only achieve a set when exposed to the air, whereas hydraulic limes, which are made from naturally occurring limestones containing 'impurities' in the form of clay minerals, will set under water in the same way as Portland cement. Readers requiring more information on building limes are referred to Holmes and Wingate (1997). Bitumen stabilisation has not been used in Britain but is widely used in the south-west USA in the manufacture of adobe blocks, where it is known as asphaltic emulsion. It does not increase compressive strength and is, in effect, a waterproofing agent, used mainly in clayey sand soils, which cements the soil particles together while at the same time coating them with a water-repellent film, effectively blocking the soil voids. Again, this is probably a non-reversible process. Although the use of soil stabilisation may be justified in certain cases, for example to improve the performance and durability of earth-based renders, it is, at least in a European context, largely unnecessary, and inappropriate in environmental terms. It has been included here mainly in order to acquaint the reader with some aspects of current earth-building practice outside Britain, and will not be discussed further because, as will be shown below, the strength and durability of earth walls can usually be improved satisfactorily by other less drastic and more environmentally sensitive means.

Modifying soils

In order to determine whether or not a subsoil requires modifying so as to make it suitable for a particular type of construction, it is first necessary to define the optimum soil characteristics for each building method. It is possible to divide load-bearing earth wall construction into two main types. First are those where clay-rich soil, to which organic fibrous material (usually straw) has been added, is mixed and placed in a wet state, either directly on to the wall in monolithic form (cob or mudwall) or in the form of pre-dried, moulded blocks (clay lump or adobe). In the second technique, damp sandy soil, containing no fibre reinforcement, is compacted by being either rammed into formwork (shuttering) using hand or pneumatic rammers – known as dynamic compaction – or formed into blocks, using a manual or hydraulically assisted compressed soil block-making press (static compaction). In the first type of construction, which might be termed the 'wet' technique, cohesion is obtained mainly by the cementing effect of the clay matrix and, to a much lesser extent, the straw binder, while in the second, which we will call the 'compaction' technique, cohesion is achieved mainly through internal friction, or interlocking, between the soil particles, with clay playing a lesser, though still important role.

For a general guide to preferred soil characteristics one may refer to Fig. 3.1, where optimum soil curves, for both the wet and compaction methods, are shown on a simplified particle size distribution chart. For further clarification, the following points should be noted: with regard to clay content, for compaction method (B) it should be between 7 and 15 per cent, whereas for wet method (A) a range from about 10 to 25 per cent would be acceptable. In method B the lower clay content is balanced by an increase in sand-sized particles. With both methods excessive amounts of silt should be avoided, though in practice this is often difficult to achieve. Above 2mm particle size – the upper limit for coarse sand – the curves combine and show equal amounts of fine and medium gravel up to 20mm diameter. However, in practice, for both adobe and compressed soil blocks an upper limit of 10mm diameter is recommended. With adobe or clay-lump blocks the clay content is less critical because the blocks are pre-dried and therefore pre-shrunk before being used. A clay fraction in excess of 25 per cent may, therefore, be acceptable, though the material would tend to be very sticky

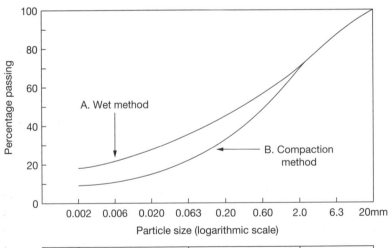

3.1
Suggested optimum particle size distribution curves for (A) cob and clay lump and (B) rammed earth and compressed soil blocks

and therefore difficult to work. A much simplified particle size distribution ratio, by percentage, of clay-silt/sand/gravel would be as follows: for wet method A, 35:35:30, and for compaction method B, 25:45:30.

In at least half of all the new-build and major reconstruction projects undertaken in the south-west of England in recent years the soils found on site have been either unsuitable or deficient in some way. As a result, considerable practical experience has been gained both in soil modification and in monitoring the performance of earth walls. Some common problems encountered with soils and how they may be dealt with are shown below:

- Excessive amounts of clay. Soils comprised entirely of clay and silt are commonly found in areas overlying the younger, Jurassic and Cretaceous geological formations in southern Britain. Such soils are quite unsuitable both for mass earth construction and the fabrication of adobe blocks. However, they can be modified quite easily by blending with either sandy soil or sand and gravel, suitably graded, obtained from a builders' merchant or quarry.

- Excessive amounts of coarse gravel and stone can be dealt with by simply passing dried, broken-up soil through a 10 or 20mm mesh. It should be noted that in doing this the clay fraction in the remaining soil will be increased.

- A deficiency of clay poses particular problems. It is possible to obtain china or ball clay in bagged, powder form from the clay producers. However, this is a very expensive solution and it is much better to find a local source of clay-rich soil and mix this, in the correct proportion, with the *in situ* material.

- Excessive amounts of fine, non-cohesive material, less than 0.5mm particle size. Some soils, those for example developed on the Permian and Triassic sandstones, mudstones and Keuper marls of western and midland Britain, are deficient not only in clay but also coarse sand and gravel. These potentially weak soils therefore pose particular problems, and this may be one case where, rather than trying to modify the soil (though with care this can be achieved) it might be better to bring in suitable material from elsewhere.

Having decided on a suitable soil, or blend of soils, it is recommended that some blocks or cylinders should be fabricated from the selected material so that they can be tested for compressive strength, and erosion resistance should this be considered necessary – for example, in cases where the walls are to remain unrendered. Chalk, and its modification by the addition of clay soil, has been discussed at some length in Chapter 2. Practical aspects of the storage, mixing and blending of soils are covered in the following section.

Storage of soils on site

Whether the proposed building site can be used for the storage of raw materials will, of course, depend on its size and how accessible it is. When a site is located in a town or in the centre of a village, then clearly access for heavy vehicles, together with space for the storage, preparation and mixing of soil, could all prove difficult, but mainly in cases where either of the two mass earth-building techniques (cob or rammed earth) are being employed. Adobe or compressed soil blocks are less of a problem as they can quite easily be fabricated off site. Two essential prerequisites for soil storage are: (1) a system by which the soil, or soils in the case of mixing or blending, can be contained and, at the same time, remain uncontaminated by other building materials or organic matter; and (2) some means of keeping the soil in an air-dry condition. It is much easier to add water to soil than to remove it. Wet soil cannot easily be broken down into its constituent parts, neither can it be sieved or effectively blended with other soils or aggregates. Storage of soil intended for cob construction in unmodified form presents less of a problem, especially if it is a free-draining sandy loam, and may, provided it is contained, be safely stored in the open and only covered when continuous heavy rain is expected. The storage area should be easily accessible to vehicles such as tipper lorries, tractors, diggers and dumper trucks.

Soil preparation and mixing

Unlike ready-mix concrete and other 'conventional' building materials – fired bricks and concrete blocks, for example, where processing is carried out by the manufacturer – earth for building cannot be purchased from a builders' merchant in ready-to-use form. Moreover, the processing and preparation of large quantities of soil by hand is both physically demanding and time-consuming. It is tempting, therefore, to look for ways in which the crushing, sieving and mixing of the raw material can be wholly or partly mechanised, thus speeding up the production process and reducing labour costs. It is possible to obtain mobile crushing, sieving and mixing machines, electrically or diesel powered, mainly from France and Belgium at a cost of around £2,500 to £4,000 apiece. However, these machines have usually been designed mainly for use in the production of stabilised compressed soil blocks and may not be entirely suitable for other construction techniques. A roller pan mixer (mortar mill) could be useful for crushing either dry soil or chalk. Upper Chalk, when dug or quarried, will always need to be broken down so that no particle is larger than about 20 to 25mm diameter because it is thought that, for both chalk cob and *pisé* construction, reducing the particle size will have the effect of increasing cohesion. Heavy-duty rotavators with power-driven wheels have been used quite successfully for breaking down soils, and large sieves can be made up from metal mesh of the required aperture size mounted in a stout timber box frame. Appropriate mixing techniques for the various construction methods, including the fabrication of earth blocks, are described below.

The terms 'wet' and 'compaction' method, which have been introduced in order to draw a distinction between the two principal ways of achieving cohesion in soils used for building, are applied to both mass (monolithic) and small unit construction. As already noted, soil preparation, blending and mixing are all made very much easier when the soil is in a dry or slightly damp condition, as saturated soil is quite impossible to work with. In all four of the principal earth construction techniques described, obtaining the correct moisture content at the mixing stage is of crucial importance, particularly in the case of rammed earth and the fabrication of compressed soil blocks (compaction method).

The use of organic fibre reinforcement

Fibre reinforcement in earth walls and mud bricks is a very ancient practice, recorded examples of which have been dated to around 6000 BC in the Middle East. In most of the traditional earth-building methods described in this chapter, with the exception of rammed earth, rammed chalk and some regional variants of cob, wet earth is mixed with organic fibre, usually straw, before being either placed on the wall or formed into blocks. In the past, various fibres were used in earth construction, including wheat and barley straw, rye straw and heather. However, in recent years the material most commonly employed has been winter barley straw, preferred to wheat straw because it has similar tensile strength but greater suppleness.

Thousands of tonnes of surplus wheat, barley, rye and flax straw are produced every year in Britain, much of which was formerly burned but which now has to be disposed of in other ways. In recent years some buildings have been constructed using compressed straw bales, a technique developed in the American mid-west, and lightweight clay/straw panels, a German innovation, are being used to form super-insulated non-load-bearing walls. Both these techniques, though interesting and relevant in the context of 'green' building, tend to fall outside the scope of this book except in cases where they may be used as outer insulation layers to load-bearing earth walls.

Fibre (straw) reinforcement in earth walls has three main functions. The first, which is particularly relevant to mass cob construction, is that it acts as a binder, allowing the wet earth to hang together in clumps or 'clats' so that it can be placed on the wall without disaggregating. The second and perhaps most important function of straw is that it distributes shrinkage cracking throughout the wall mass, preventing the development of serious fissures at key points in the structure (see Fig. 3.2). In addition, it is thought that individual fibres of long straw may act as conduits, channelling water

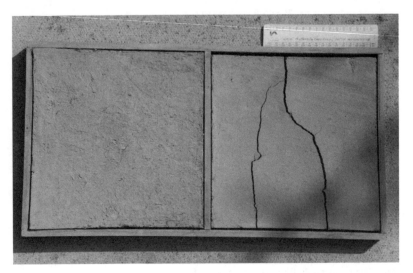

3.2
Controlling shrinkage by the addition of organic fibres. The soil on the right is fibre-free

from the core of the wall to its outer face, thus accelerating the drying-out process. The third function relates to tensile strength, which, according to research carried out at Plymouth University, is significantly increased when straw is added to the soil (Saxton, 1995). This research was carried out using newly fabricated specimens rather than material from existing cob walls. An additional function of straw, if large amounts are added to the soil (up to around 2.5 per cent by weight), is to significantly reduce its density, by between 10 and 12 per cent, thus also reducing its thermal conductivity. Although intact straw has been found in demolished cob walls two or three hundred years old, it is sometimes the case that, especially in walls that have been badly affected by dampness, the fibres will have almost entirely decayed. In cob construction long straw, straight from the bale, is to be preferred for the reasons outlined above. In the fabrication of adobe blocks or mud bricks, on the other hand, it may be more convenient to use chopped straw, cut into 100mm lengths.

Wet methods – cob, cob blocks, clay lump and adobe

For all the wet methods the mixing technique is basically the same. The principal aim is to ensure that the clay and fine silt particles, known as the 'binder fraction' because of their cohesive properties, are distributed evenly throughout the remaining, coarser particles – the 'aggregate fraction'. Maximum cohesion, and therefore optimum tensile and compressive strength, is achieved only when all the coarser aggregates are entirely contained within a clay matrix. In Chapter 2 the terms 'dispersion' and 'flocculation' were introduced to explain how, in general terms, excess water in a soil will force the clay particles apart so that they are, in effect, in suspension; and that during subsequent drying out the inter-particle bonds are re-established in such a way that optimum distribution of the clay fraction is achieved.

Cob, which may be described as wet mixed and placed, fibre-reinforced mass subsoil construction, represents the mainstream British earth-building tradition. It is estimated that at least 75 per cent of all surviving earth buildings in the United Kingdom and Ireland were built using this technique, and similar buildings may be found in northern France, eastern Germany, Hungary, the Czech and Slovak Republics, and in other parts of Europe. Although cob building in its traditional form is unlikely ever to be adopted for volume house building, it is a form of construction ideally suited to the self-builder or to environmentally concerned housing communes. One reason for this is that the tools and equipment required for mixing cob and for the fabrication of adobe and clay-lump blocks are either simple, inexpensive hand tools or plant and machinery that can be hired easily and at relatively low cost. The basic mixing procedure requires the soil and straw to be compressed (simulating the effect of treading, either by foot or the hooves of cattle – both traditional methods), thoroughly wetted in a progressive and controlled fashion, and periodically scraped up and turned (see Fig. 3.3).

3.3
Traditional method of mixing and placing cob

Attempts have been made to mix cob by using machines such as tilting-drum (concrete) mixers, roller pan mixers (mortar mills) and rotavators, but none of these have been entirely successful because no one method has proved to be capable of performing all three functions in one operation.

Cob mixing

The revival of traditional earth building in Britain has its origins in Devon. At the beginning of the 1980s Alfred Howard, a craftsman builder living and working in the Crediton area, started building in cob using largely forgotten traditional techniques. The mixing method described below was developed by Mr Howard over 15 years ago and continues in use to this day.

The key item of equipment required is a tractor fitted with a front loader/scraper bucket or a mechanical excavator, known more commonly as a digger, or a backhoe in the USA. Wheeled rather than tracked vehicles are to be preferred because the latter are rather slow moving and difficult to manoeuvre in this situation. In the USA and Australia the 'Bobcat' skid steer loader is extensively used for the transporting and placing of soil in rammed earth construction projects. This type of machine may also be suitable for cob construction, though the author is not aware of it having been used for this purpose. For mixing by machine a flat, level, hard-surfaced area about seven by four metres, with additional space for a tractor to turn at one end, is considered most suitable (see Fig. 3.4). Apart from the tractor driver, at least two, preferably three, other persons are required. Also essential is a hose, fitted with a spray head, connected to a mains water supply. With regard to the other essential ingredient, straw, which for cob is used straight from the bale rather than being chopped, experience has shown that about 1.0 to 1.5 per cent, by weight, gives best results with sandy soils, though clay-rich soils may need more in order to soak up excess clay and water. When more than around 2.5 per cent straw is added the material becomes unworkable. Straw bales can vary somewhat, both in size and degree of compaction, so assessing the volume of straw to be added is largely a matter of experience. As a general guide, one bale of straw (weighing around 25 to 29kg) added to 1.5 tonnes of loose, dry soil (volume about 1.0m^3) should be about right. The amount of water required to bring the soil to its ideal consistency will be dependent mainly on the size of its clay fraction. Clay-rich, silty soils will take up much more water than, for example, coarse, sandy soils. Thorough mixing requires the water content of the soil to be significantly higher than it would be for placement on the wall. A state of optimum consistency has been reached when all the straw has been entirely coated with clay and assumes the same colour as the soil.

The mixing area is first well wetted and then completely covered in a thin layer of straw. The tractor is used to fetch dry or slightly damp soil from

3.4

**Mixing cob using
a mechanical
digger**

3.5

**Cob drying out on
timber pallets**

the storage bay, which is spread out in a layer 150 to 225mm deep over the straw. When this has been done a layer of straw, about 100 to 150mm thick, is added and the hose used to thoroughly dampen the mix. When sufficient water has been added the tractor is driven back and forth in such a way that its wheels force the straw into the soil. While this is going on, the person supervising the work will periodically check samples of the material in order to decide whether more straw or water, or possibly both, need to be added. By this time the bed of cob will have been thoroughly compressed, and the tractor bucket is used to scrape up the material and turn it over. The mixing procedure is then repeated, more straw and water being added as and when necessary. Tradition has it that cob should be thrice turned and trodden. There is little doubt that, as with lime mortar, the more the material is mixed the more its consistency and workability are improved. When the mixing process is complete the cob will usually be too wet for immediate use. It should be heaped up, preferably on timber pallets (see Fig. 3.5) or raised decking, and allowed to stand, overnight in warm, windy conditions, longer if necessary, until it has attained a fairly stiff but workable consistency; at this point it will be ready for placing on the wall. By using this method it should be possible, using a digger, to mix around five tonnes of cob in one-and-a-half hours. For anyone new to cob building it is recommended that, in order to become familiar with the finer points of cob mixing, it is very helpful to first mix a small batch – one or two tonnes – using the manual method described below.

It should perhaps be noted that, until quite recently, cob was usually mixed to a consistency at which it would be ready for immediate use, so that it could be mixed and then placed on the wall on the same day. Clearly, in terms of time and labour costs, this makes a great deal of sense. However, experience seemed to indicate that unintentional over-wetting of the material, a not infrequent occurrence, followed by a period of drying out, had the effect of significantly improving the homogeneity and workability of the cob. That its long-term strength and durability may also be improved in this way has subsequently been confirmed by research carried out at Kassel University in Germany (Minke, 2000). It should also be noted that chalk reacts to water in a different way to clay-based soils, so that adding water to chalk needs to be done in a carefully controlled way, and it may be that the manual method of mixing should be adopted, even in the case of chalk cob where a mix of chalk and clay soil is used. The traditional method of using animals to mix cob, oxen in former times, now more usually bullocks, should be mentioned in passing. The cloven hooves of cattle are ideally suited to the treading of cob and, in some recent cases where buildings have been constructed on or near cattle farms this method has been adopted with some success. However, it is still a very labour-intensive and time-consuming process compared with machine mixing.

Mixing procedure for adobe, clay lump and cob blocks

Unfired mud bricks probably represent the oldest known 'mass-produced' building material. Adobe is the Spanish form of an Arabic word meaning brick, and has now become a generic term used to describe all sun- or air-dried mud bricks. Adobes are easy to make using simple hand tools, and in the warmer, more arid parts of the world, have been used for at least 7,000 years to construct buildings of considerable size. The basic ingredients for mud bricks or clay lumps (the East Anglian version of adobe) are much the same as those for mass cob. The principal differences concern stone and clay content, the use of chopped straw and a manual mixing procedure. As for any form of small unit masonry construction, the dimensions of a mud brick or block are determined not only by wall thickness but also by the size and weight that may be comfortably handled by the mason. The most usual size, internationally (adopted as a UN standard) is about 300 × 150 × 100mm, though they can be fabricated to any desired dimensions. For example, in south-west England, blocks the same size as the standard concrete item, 450 × 225 × 100mm, have been extensively used in major repair works, while in East Anglia clay lumps up to 450 × 225 × 225mm (weighing 36 to 38kg each!) have been noted. Large cob and adobe blocks, when laid on flat rather than on edge, make for a very strong, durable wall.

When preparing the soil for block making, all stones larger than about 10 to 15mm should be removed by sieving or screening and the straw should be cut to a length of between 100 and 150mm. For mixing small batches of clay-based cob or for chalk cob the straw can be taken straight from the bale. The traditional mixing procedure, which involves treading the soil and turning it with a fork, can be carried out in a relatively small area. A shallow timber box about 1.2 × 1.9m with sides 200mm deep would probably be adequate if no more than two people were mixing. With manual mixing the sequence of events is much the same as for machine mixing, the principal difference being that the process is more thorough and can be carefully controlled. Layers of soil should be no more than 75 to 100mm, a depth that can be penetrated effectively by the boots of the workers. In 1994, during the course of major rebuilding works to a historically important cob farmstead near Lapford in mid-Devon, around 20 tonnes of material was mixed manually, and in Fig. 3.5 the cob is shown drying out on pallets prior to being placed on the wall.

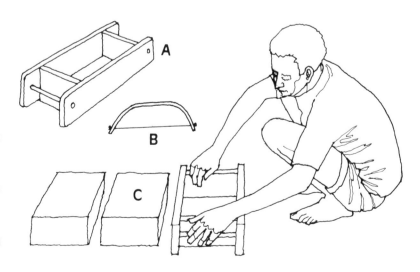

3.6
Traditional method of making mud bricks (adobes) showing (A) typical bottomless mould box, (B) wire for trimming off surplus material and (C) completed bricks

with renderings

Fabrication of mud bricks and cob blocks

The principal difference between mud bricks (adobe and clay lump) and cob blocks is that the former are made using wet material and the latter uses a somewhat stiffer mix, of a consistency similar to that of mass cob when it is placed on the wall. In the Middle East and North Africa a simple timber mould is employed (see Fig. 3.6). A lump of wet soil and straw is thrown into the mould, completely filling it. A wire or plasterer's float is then drawn across the top of the mould to remove any excess material and the mould is lifted off. Experienced workers can make 300 to 400 adobes in one day. East Anglian clay lumps are made in much the same way, although the daily output is much lower due to the larger size of the blocks. As soon as they are dry enough to handle, the blocks are stored in a covered but well-ventilated area, being turned periodically, until they are thoroughly dry and ready for use. Traditional adobes should not be confused with the adobe blocks currently produced in the south-west USA. These are essentially soil blocks, often stabilised with cement or bitumen and containing no fibrous reinforcement. They are produced using a highly mechanised, carefully controlled manufacturing process, which enables many thousands of blocks to be produced in one day. The hot, dry climate of this region allows the blocks to be fabricated, dried and stored in vast outdoor 'adobe yards'.

The fabrication of cob blocks has become something of a cottage industry in the south-west of England, where many thousands of them have been used to carry out major structural repairs to existing buildings. The most commonly used block is the same size as the standard four-inch concrete item – 450 × 225 × 100mm, though 'quarter blocks' (225 × 112 × 100mm) not much larger than a fired brick, have also been used for minor repairs (see Chapter 7). The type of mould most usually employed is a bottomless ladder or gang mould, in which four large blocks can be made in one operation. Making good blocks in this way is an acquired skill and the process itself is rather time-consuming and labour intensive. Two men, using ready-mixed cob, can produce between 100 and 125 large blocks in one day using this method. The earth/chopped straw mix is prepared in the same way as mass cob, using either the machine or manual method. The material must be wet enough to allow the mould box to be filled without excessive treading or ramming, but stiff enough to prevent the blocks deforming or 'slumping' when the mould is removed (see Fig. 3.7). An alternative method is to use a concrete block-making machine. These were manufactured during the period following the end of World War II, when building materials and plant were in short supply. A few still survive and have been adapted for use with cob. Only one block can be made at a time, so the production rate is probably no faster than that for the gang mould method described above. Compressed soil block presses,

3.7
Above, cob blocks
(450 × 225 × 100mm)
fabricated using a
four-gang timber
mould and, below,
blocks stored for
drying out, also
showing gang-
moulds

which are described in the following section, have also been employed and can produce 40 to 50 medium-sized blocks in one hour.

Compaction methods – rammed earth (*pisé*) and compressed soil blocks

As already noted above and in Chapter 2, cohesion as a result of close packing, or interlocking, of the soil particles can be achieved by means of compaction. It has also been noted that heavy compaction of soil can result in a 10 to 12 per cent increase in its dry density. Whereas in the wet method the soil is mixed at a high moisture content in order to achieve maximum cohesion, in the compaction method, where the principal aim is to achieve maximum dry density, a much smaller amount of water is added to bring the soil to what is known as its optimum moisture content or OMC.

Rammed earth

One American author, David Easton, has described the technique thus:

> Think of rammed earth as a sort of 'instant rock'. The natural process of creating sedimentary rock occurs over a span of thousands and millions of years. An earth rammer, on the other hand, creates it in a matter of minutes.
>
> (Easton, 1996)

It is not surprising that in Australia the technique is known as 'engineered earth'. Much research has been devoted to developing rammed earth techniques over the past 70 years, mainly in the USA, and as a result of this two main criteria have emerged. The first, which has been referred to above, relates to particle size distribution – a preference for soils with a high sand and low clay content, and the second, which is critical to the mixing process, concerns moisture content. The optimum moisture content (OMC) is the moisture content at which maximum dry density is achieved. In other words, only sufficient water to act as a lubricating film between individual particles, enabling them to move freely in relation to one another, is added to the soil.

OMC is determined by what is known as the Proctor test, in which soil at various moisture contents is tamped into a steel cylinder of given volume in three layers of equal depth until maximum density is achieved (this test is fully described in BS 1377, Part 4). The Proctor test is also considered suitable for determining the OMC for soil to be used for the fabrication of compressed soil blocks. However, at least two authors (Vargas Neumann, 1993; Minke, 2000) have suggested that the results of this test may not be totally reliable, that they represent the minimum rather than the optimum moisture content, and that both compressive and tensile strength may be improved by increasing the moisture content by up to 20 per cent of the measured OMC. If the soil is air-dry its moisture content will be around 3 per cent, so that if a drum or pan mixer is used it should be possible to add the required amount of water, by weight rather than by volume, in a controlled way using a watering can fitted with a fine-spray head. For larger amounts of soil, mixed using a Bobcat or tractor, the material would need to be periodically checked by an experienced builder. For anyone contemplating rammed earth construction it is recommended that expert advice should be sought from a specialist contractor or architect experienced in this form of construction.

Compressed soil blocks

The difference between dynamic and static compaction has been mentioned above. It is considered that the repeated heavy blows and vibration resulting from manual or, more particularly, pneumatic ramming is more effective in achieving 'densification' in soils than is the 'one-off' static compaction applied by a manual or hydraulically assisted block-making press. The standard CINVA-ram-type manual block press, which was first developed in Bogota, Colombia, and has been around in various forms for some 50 years, was designed for use with cement-stabilised soils and was intended for the construction of low-cost housing in developing countries (see Figs 3.8 and 3.9). It is simple to operate and can produce around 500 blocks per day. The soil-mixing procedure is the same as that for rammed earth, apart from the need to screen out stones and gravel over 10mm diameter. When used with unstabilised soil the manually operated machines will produce blocks having a compressive strength no greater, sometimes less, than that of an adobe or clay lump (Minke, 2000). The principal advantages of compressed soil blocks are that they are of geometrically precise, uniform dimensions and, when released from the mould box, can be handled immediately and are ready for use after storage for one week in dry, well-ventilated conditions.

Foundations and plinths (underpin courses)

In most traditional earth walls the plinth also formed the foundation, or footing. However, for new earth construction, in order to comply with Part A of the Building Regulations, some form of separate foundation will normally be required. Most mass earth walls will have a width of 450 to 600mm at their base. The width of traditionally built cob walls in rural areas, most of which were probably built without the aid of scaffolding, was determined by the need to accommodate a workman placing and then treading the material on the wall head. Rammed earth walls, because of their higher strength and bearing capacity, need not be as wide as cob walls, usually 450mm or less, though walls 550 to 600mm wide are common in the south-west USA. It should be stressed that, at this stage, we are considering earth walls only in terms of their structural stability and load-bearing capacity. If, in order to comply with the thermal performance requirements of Part L1 of the Building Regulations, additional layers of lightweight material need to be added to either the internal or, more likely, external wall face of the wall, then clearly the width of the underpin course, or stem, as it is known in the USA, will need to be increased. This and other issues relating to Building Regulations and Codes are discussed in much greater detail in Chapter 5.

In most recent new-build cob projects a conventional horizontal concrete strip foundation has been employed. Generally speaking, a two-storey cob house would require a concrete base, 300mm wider than the

3.8
CINVA ram-type
compressed soil
block press,
manufactured in
Belgium by
UNATA

width of its walls with a minimum depth of 150mm. It has been estimated that a house of this type, assuming a 750mm-wide foundation, would exert a ground-bearing pressure of around 90kN/m^2. For those wishing to use recycled materials, well-compacted salvaged road scrapings could prove suitable, subject of course to the approval of the Building Control Officer. It should be noted in this connection that, provided the underlying subsoils are stable, totally rigid foundations may not be essential for mass earth walls because, unlike conventional (modern) masonry walls, they can tolerate a certain amount of minor movement without developing serious cracks and fissures.

The principal functions of the masonry plinth, or underpin course, are (1) to provide a firm, level base to support the earth wall and all its imposed loads, and (2) to prevent moisture, in the form of rising or penetrating dampness, from adversely affecting the base of the earth wall. In order to function

3.9
**Operating a
manual soil block
press**

effectively the top of the plinth should be a minimum of 450mm above external ground level. In most parts of Britain local stone would have been the
material employed, though fired bricks were sometimes used, in East Anglia
for example. Poorly constructed plinths, often having no cross-bonding, and
with their cores filled with rubble and earth, are a frequent cause of problems
in old cob buildings (see Chapter 6). Traditionally, damp-proof courses (dpcs)
were never incorporated in plinth walls. This is because in 'breathable' construction systems they were, and still are, not strictly necessary, as excess
moisture is able to evaporate freely through the porous fabric of the building
(see Chapter 6, Fig. 6.4). However, compliance with Part C of the Building
Regulations will normally require the installation of a dpc and this must be
planned for at the design stage, and when selecting materials.

Embodied energy input is much reduced if building materials can
be sourced locally. If, for aesthetic reasons, a stone plinth is preferred, it may
be possible to find salvaged material or acquire, from a local quarry, random
stone that is normally crushed for road construction. Salvaged bricks may also
be available locally. In order to achieve the low thermal transmittance (U-value)
now demanded by British and other European building codes, an inner leaf of
low-density blocks is often incorporated in the plinth, the top of which may
be raised to window cill level in order to reduce the overall U-value of the wall.
These matters are dealt with more fully in Chapter 5. If a dpc is to be installed,
it should be located within the plinth, at least 225mm below the first lift, or
course, of earth. The reason for this being that if excess moisture, resulting

from driven rain or flash flooding for example, is prevented from dispersing by means of evaporation or capillary movement, it may seriously weaken the wall at the point where it is most vulnerable to failure. For this reason, the plinth should never be rendered, but always left exposed. How the upper face of the plinth is finished will depend on the type of earth construction to be employed. For cob construction the plinth needs to be approximately level and provision needs to be made to ensure that the first lift of cob is keyed into the masonry in order to prevent it moving, both while under construction and when drying out. This can be achieved by either leaving stones, or bricks, projecting vertically from the centre of the wall every 500mm or so, or by leaving a 150 to 200mm deep groove, also in the centre of the wall. In the case of rammed earth the plinth should be flat and level. Depending on the type of shuttering to be used, putlog channels, or grooves, across the full width of the wall, may need to be constructed (see Chapter 4). For adobe, cob block or compressed soil block construction the plinth needs only to be flat and level.

On-site access and safety

Height from ground level to eaves in a two-storey domestic building is going to be in the region of 4.5 to 5.0 metres. In order to comply with current British health and safety regulations, any building work carried out at or above a height of 1.5 metres will need to be performed from a stable working plat-form supported on scaffolding. Unfortunately, in the case of mass earth construction, the need to provide a safe working environment can restrict working space and limit accessibility, which is why, in the past, some working practices that might now be considered somewhat hazardous were fre-quently adopted. Lifting large quantities of damp earth to a height of three metres or more can pose particular problems, even when the raw material is being mixed close to the building. However, these problems can, to a certain extent, be overcome. That most useful machine, the tractor, for example, if equipped with a long reach front-loader bucket and a rear forklift, can place the raw material where it is needed, either on to external working platforms or inside the building. If, as work proceeds, the ground and first floors are finished to the point where they can provide a stable, level working area, then this will also speed up the building process.

Chapter 4

Earth construction 2: the building process

By L. Keefe and R. Nother

Cob wall construction

As noted in the previous chapter, the ideal consistency of cob for placement on the wall is when the material has attained a fairly stiff but plastic state, somewhat like dough or putty. It should hang together in fairly large clods that can be easily lifted with a fork. If the cob disaggregates into small pieces it is too dry and would need to be wetted and re-mixed. As a general guide, a lift of cob should be equal to the width of the wall, though it is possible to construct higher lifts, up to one metre, if coarse, sandy soils are being employed.

Tools and equipment

For placing cob on the wall, suitable footwear – stout, heavy-duty boots – are essential, as are shovels and forks. Purists prefer to use the traditional dung fork, perhaps because of its longer reach, though standard garden forks with sharp, narrow tines work almost as well. Also useful are tools for beating the cob into place. Heavy wooden mallets and lead-dressing tools are commonly used (see Fig. 4.1) though some builders prefer to make their own 'cob beaters': a square piece of oak about 450 × 100 × 100mm with a 450mm-long handle, for example. For removing surplus cob and paring down the wall face, a medium-sized mattock and/or a spade with a flat, sharp-edged blade are commonly employed.

Forming door and window openings

Before placing the cob, provision must be made for the incorporation of door and window openings. Ensuring that the walls are built vertical and square can be achieved by means of periodic monitoring as construction proceeds (this topic is further discussed below). It is, of course, possible to build circular, curved or elliptical walls in cob, or 'battered' walls that taper from ground to eaves level, because the material is so flexible and accommodating. It is all a matter of careful setting out, and control during the building process. It is claimed that, in the past, cob walls were constructed in totally monolithic form, but with lintels built into the wall so that door and window openings could be cut out when the cob had dried. Unlikely as it may seem, there is no doubt that some buildings were constructed in this way. Today, however, it has become normal practice to create openings using robust timber formwork with sufficient cross-bracing to resist the lateral pressure resulting from treading and beating the cob into place (see Figs 4.2 and 4.5).

Placing the cob

A typical building procedure would be as follows. Two builders stand on the ground, or on a suspended working platform, turning and re-treading the cob as necessary, and placing it on the wall head for a third person to tread and beat it into shape. Treading the cob will cause it to spread, but by no more than 100 to 150mm. If it spreads more than this, the material is too wet and should be left to dry out for a while. Beating the material back, either from above or from the side, will help to contain it and ensure that good compaction is achieved. As well as the tools mentioned above, the fork, or the heel and sides of the builder's boot are quite effective in containing and compacting the cob (see Fig. 4.3).

Particular care should be taken to ensure that the material has been well compacted at the corners of the building as these are the areas most

4.1
Placing cob using a dung fork and lead dressing tool

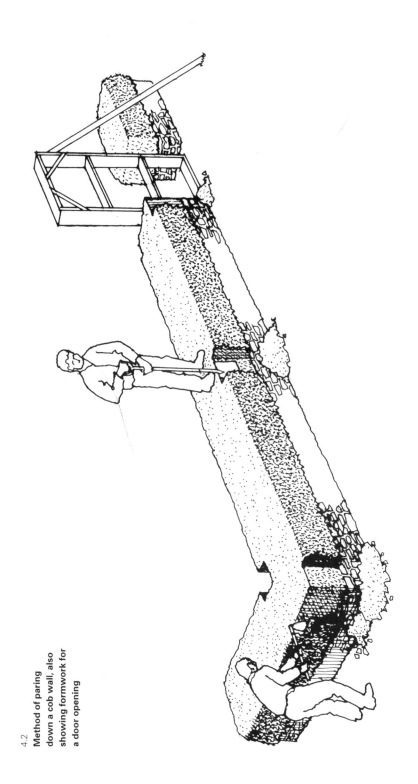

4.2
Method of paring down a cob wall, also showing formwork for a door opening

susceptible to damage through abrasion or weathering at a later stage. The internal corners are also the points at which shrinkage will be concentrated during the drying-out period. When the top of the first lift, usually about 600mm above the plinth, has been reached the material is left overnight to start drying out and, if rain is expected, covered with bales of straw or corrugated iron sheeting. Some soils, moderately clayey medium to coarse sands, for example, will dry out much faster than others. In warm, windy conditions the first lift of cob may have consolidated sufficiently for the next lift to be placed within two to three days. Full consolidation has been achieved if it is possible for a couple of people to jump up and down on the wall top without it slumping or spreading. Whether or not the material is fully consolidated, as long as it can support the weight of a person without deforming, excess cob should be removed by paring back to the finished wall face before the material becomes too dry and therefore difficult to remove. Surplus material can be kept back and incorporated into the next batch of cob when it is mixed.

The walls of many older cob houses, in particular those in rural areas, are neither truly vertical nor square. Rounded corners, gently undulating wall surfaces and door and window openings that are out of square all contribute to their picturesque charm and idiosyncratic character. It is, of course, perfectly possible to create 'organic' or neo-vernacular buildings, which exploit the sculptural qualities and aesthetic potential of the material rather than being simply functional, and some of these are illustrated in Fig. 4.4 and Colour Plates 3 and 4.

4.3
Treading and beating cob

However, assuming that the walls of most contemporary buildings will need to be square and vertical, one should proceed as follows. The external and internal corners are first squared off, which can be done by setting up a string line from vertical posts placed at each end of the wall, running parallel to the finished wall face. Where the string lines of two adjoining walls intersect will locate the corner of the plinth. Alternatively one can carefully pare down the corner sections using a mattock and spirit level or plumb line, or use vertical formwork, as shown in Fig. 4.5. Additional vertical slots can then be cut out at intervals along the wall, as shown in Fig. 4.2. One then

4.4
Curved cob walls under construction. The completed building is shown in Colour Plate 4

4.5
Cob wall under construction, showing timber formwork in position at corners, doors and window openings

has an accurate line to work from, which will ensure that the first two or three lifts of cob are constructed vertical and true, by which time scaffolding or some form of working platform will need to be put in place.

A note on shuttered cob

At this point it has probably occurred to the reader that a great deal of time and effort might be saved by simply treading or ramming the cob into shuttering. In fact, there is documentary and anecdotal evidence to show that, at least in Devon, this method was employed during the nineteenth century, from around 1820 right up to c. 1914, certainly for agricultural buildings, possibly also for town houses and villas. Apart from the need to constantly dismantle, move and re-assemble the formwork, shuttered cob has other disadvantages. Wet cob – as employed in the 'puddled clay' method – will, because it is contained behind shuttering, be very slow to dry out, and it would appear that, in the past, some shuttered walls were, in order to speed up the drying process, constructed with material containing insufficient water. One fairly obvious result of this is that the wall face, once the shuttering is removed, can be seen to contain numerous cavities (see Fig. 4.6). The other result, less obvious, will be that a wall constructed in this way might tend to be weaker than one built using the traditional method because the material has been insufficiently compacted. Of course, the appearance of a wall may be of little concern in cases where it is intended to apply an earth- or lime-based render, or where an external insulating layer or some form of cladding is to be applied. Shuttered cob has been employed successfully, in conjunction with 'piled' cob, as part of major reconstruction works carried out at two historic sites, both in Devon: Bowhill House, Exeter, and Bury Barton Farm, Lapford, mid-Devon. In both cases the material was placed at the same moisture content as that of piled, unshuttered cob. Horizontal scaffold planks were used as formwork, securely fixed and wedged to tubular steel scaffold poles, and the wall faces pared down after removal of the forms. This system has also been used for a new-build project at the Norden Visitor Centre, Dorset, in 1998 (see Figs 4.7 and 4.8 and Colour Plate 6).

Placing the cob (continued) – fixings and structural supports

Before placing the second lift of cob, one needs to give some thought to fixings for door and window frames. Nails and screws are not suitable, neither is the insertion of fibre or plastic wall plugs into pre-drilled holes, although iron spikes and hardwood pegs can be driven into cob when it is still in a damp, semi-solid state. Helifix stainless steel helical wall ties can be used to obtain a secure fixing into dry cob, but these are used mainly in repair situations (see Chapter 7). Oak is well known for its durability, high strength-to-weight ratio and resistance to damp, and is therefore well suited for structural use in cob

4.6
Exposed shuttered cob in the 430mm thick wall of a barn in mid-Devon, built _c._ 1850. Also showing a half-brick facing applied to a badly eroded west-facing wall _c._ 1935

4.7
Upper section of rebuilt wall under construction using shuttered cob

4.8
**New wall built
using shuttered
cob. The
completed
building is shown
in Colour Plate 6**

buildings. Current practice, which is closely based upon historical precedent
and was originally developed by Alfred Howard in Devon, is to use oak for
fixings as well as for structural elements such as bearer/spreader plates,
lintels and wall plates. Fixings for door and window openings are bedded and
pegged into the cob as work proceeds. They can be butted up against the
formwork so that they are secured in position while the cob dries out (see
Fig. 4.9). Fixings for window cills can be cut to a dovetail section in order to
secure them into the cob. Lintels must be supported on spreader plates,
pegged into the cob, that extend across the full width of the wall and for at
least 225mm either side of the door or window opening. The size of lintels,
of which there are normally three for a wall thickness of 550 to 600mm, will
be determined by the width of the opening. Typically, for an opening of up to
1,200mm, there would be two outer lintels of 225mm wide by 150mm deep
and an inner lintel of 100 × 150mm. In some repair works to cob buildings
concrete lintels, supported on slates bedded in cement or hydraulic lime
mortar, have been used to replace decayed or damaged timber lintels. The
justification for this being that, in purely practical terms, concrete is much
cheaper, and possibly more reliable, than seasoned oak. If circular or arched
openings are to form part of the design, these can quite easily be constructed
using formwork. A round or elliptical opening, or a narrow pointed (Gothic)
arch would not need a lintel. This is because the monolithic nature of mass

4.9

**Detail of window
opening in cob
wall**

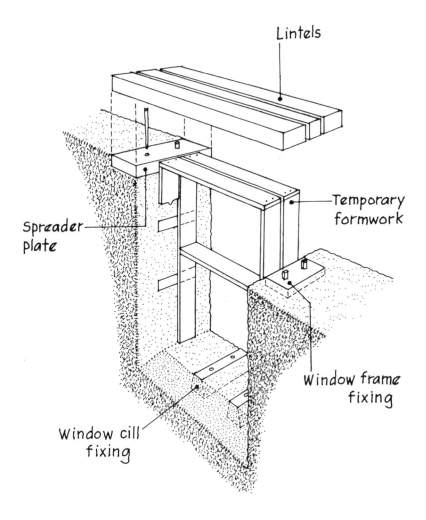

Lintels

Temporary
formwork

Spreader
plate

Window frame
fixing

Window cill
fixing

cob means that it is largely self-supporting, in much the same way as mass
concrete. However, a shallow arch would require a lintel to be placed above
its highest point and would probably need to be constructed using either earth
bricks or timber framing covered with an earth plaster.

The first floor, assuming there is one, will need to be supported
off the cob walls. In many traditional buildings, the floor joists were notched
into large oak cross-beams supported on bearer plates and, at the gable walls,
simply bedded in the cob. Normally, no provision was made for corner reinforcement in the cob at first-floor level. In contemporary domestic buildings
it is customary to support floors and ceilings using treated softwood joists
only. If this system were used it would be necessary to ensure that the joist
ends were treated with a timber preservative to prevent problems that might
arise from penetrating dampness. Building support plates into the wall, to
which the joists could be fixed, would probably be a good idea as these could

be continued around the corners in order to provide additional reinforcement (see Fig. 4.10). Internal load-bearing walls can be constructed using 300 × 150 × 100mm earth blocks or, if non-load-bearing, lightweight earth blocks of the same size. Earth block construction is discussed below.

At eaves level, wall plates may be spiked or pegged directly into the cob, or screwed into dove-tailed oak cross-pieces, which are themselves pegged into the wall head (see Fig. 4.11). The wall plate can be extended round the corner of the building into the gable wall, as at first-floor level, in order to restrain any movement resulting from shrinkage cracking. Whether or not some form of earth construction continues above eaves level will depend on the design of the roof and gable walls. If the roof is to be fully hipped, then the wall plate will continue around the perimeter of the building in the form of a ring beam (bond beam in the USA). The traditional fully hipped, thatched roof was ideally suited to cob buildings because the deep eaves overhang provided good protection from the weather. To form a half-hipped roof the cob can be built up to the required level, with steps or channels formed in the wall head to support purlins and rafters. Building up to the apex of a full-height gable wall using mass cob is difficult, though quite possible if scaffolding is used and the cob is tamped into shuttering rather than being piled in the usual way. In the past, the section of the gable wall between eaves level and apex was often constructed using some other material –

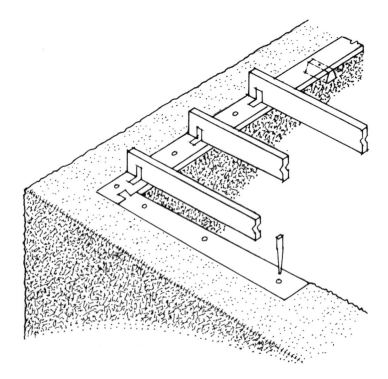

4.10
Detail of first-floor joist support plate and corner reinforcement

timber framing with lath-and-plaster and weatherboarding or slate hanging, or nine-inch (225mm) fired brick. On balance, the half-hipped roof may represent a compromise solution or, if a full-height gable wall is required, building up to the apex using 300mm earth blocks might, in the case of a tiled or slated roof, make construction of verge overhang, soffit and bargeboard rather easier (see Fig. 4.11). In the traditional buildings of south-west England, where the common rafters were normally bedded directly on the cob wall head rather than on a timber wall plate, the final layer of cob was taken up to the top surface of the rafters. This was, and still is, known as beam filling and its primary purpose was probably to secure and prevent lateral movement of the rafters. Ventilation of roof spaces, at a time when most cob houses had thatched or hand-riven slate roofs, was clearly not an issue.

Notes concerning shrinkage and settlement in mass earth walls

Soil expansiveness has been referred to in Chapter 2, where in Fig. 2.3 a state of soil consistency known as the shrinkage limit is shown. If a soil could be placed on the wall at this moisture content, then neither linear nor vertical shrinkage (settlement) would occur. Although this condition might be approached in rammed earth walls, where linear shrinkage (at optimum moisture content) is usually no more than 0.25 to 0.5 per cent, in the case of wet-placed cob a degree of shrinkage is unavoidable. The only exception to this is cob made from Upper Chalk, which, because it contains no clay, is non-expansive, though some settlement might occur if the chalk were not allowed to consolidate fully between lifts. Both linear shrinkage and vertical settlement would be about the same in carefully built walls, varying from less than 1 per cent in low clay, coarse sandy soils up to 3 per cent in fine silty soils containing in excess of 20 per cent clay. Although fibre reinforcement in cob has the effect of reducing overall shrinkage, this may still occur and will tend to be concentrated at openings in the walls and at the corners of the building. Internal corners are particularly vulnerable because at these points the material is pulling in two directions at an angle of 90 degrees, which is why the incorporation of some form of corner reinforcement is considered useful, especially when expansive clay soils are being employed. In a curved wall this problem would not normally arise. Curved, and especially circular, walls are much more resistant to damage and structural failure than are straight walls.

Fireplaces and chimney stacks

In many cob and clay-lump buildings chimney stacks were built entirely of earth, apart from the fireback, which was normally lined with fired brick or masonry. Mass earth is regarded, for Building Regulation purposes, as a non-combustible material and, contrary to popular belief, is not adversely affected by exposure to high temperatures. Examination of cob walls following serious

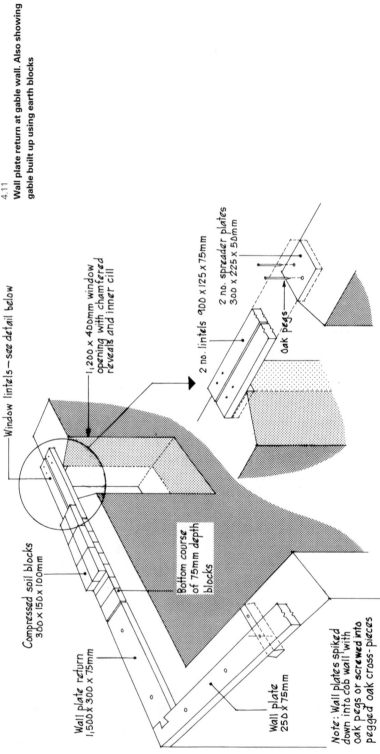

Window lintels — see detail below

1,200 x 400mm window opening with chamfered reveals and inner cill

2 no. lintels 900 x 125 x 75mm

2 no. spreader plates 300 x 225 x 50mm

2 no. lintels 900 x 125 x 75mm

Oak pegs

Compressed soil blocks 300 x 150 x 100mm

Bottom course of 75mm depth blocks

Wall plate return 1,500 x 300 x 75mm

Wall plate 250 x 75mm

Note: Wall plates spiked down into cob wall with oak pegs or screwed into pegged oak cross-pieces fires would seem to suggest that, rather like oak, damage is confined to surface charring and that the main body of the wall remains structurally sound. In fact, much more serious damage is caused by prolonged drenching with fire-hoses. Chimney stacks constructed of piled cob can, when they are built into external walls, be a source of problems for two reasons: (1) where they project forward of the external wall face they represent potentially weak points in the structure, especially when, as is often the case, they have been extended vertically by the addition of a brick shaft, and (2) being fairly exposed to the weather, they are liable to be affected by penetrating dampness (see Fig. 4.12). Axial chimney stacks, built at the centre of the house as part of a load-bearing cross wall, are usually less of a problem structurally, though still subject to dampness at ridge level and above. These

4.12
Outward leaning cob chimney stack and gable wall; the result of rising and penetrating dampness, exacerbated by the presence of an impermeable cement/sand render coat and a blocked, unventilated chimney flue

problems, together with others relating to earth chimney flues, are further discussed in Part II, Chapter 6.

The traditional fireplace and chimney was, particularly in higher status yeoman farmhouses, often a fairly massive affair, with a large oak lintel, or bressumer, supported on masonry jambs, upon which the cob chimney breast was built. Fireplaces were large because, in addition to heating, they were also used for cooking and often contained a built-in bread oven. At the bottom of the social scale, in labourers' cottages, which often had only two rooms, one up and one down, the fireplace lintel would frequently be supported on jambs built of cob. This type of basic cottage might also have a chimney breast in the form of a canted smoke hood constructed not from piled cob but with cob applied to a timber-framed, wattle hurdle arrangement, 100 to 150mm thick in total.

Building a fireplace, chimney breast and stack in compliance with the relevant Building Regulations should not present any particular problems. Ideally, the chimney breast and stack should form an integral part of the structure, being constructed from piled or shuttered cob raised at the same time as the main walls. In some recent neo-vernacular houses the outer, projecting parts of external chimney stacks have been built in stone, securely keyed into the inner cob wall. Stainless-steel fixings such as expanded metal lathing, though obviously non-traditional, are ideal for this purpose as well as for forming corner reinforcements in projecting sections of cob. Their use in achieving secure bonding and joints between earth and other building materials is discussed in Chapter 7. The experimental cottages built at Amesbury, Wiltshire, in 1919–20 by the Ministry of Agriculture and the Department of Scientific and Industrial Research, were comprehensively recorded and documented (Jaggard, 1921).

According to Jaggard's report, several of the cottages, which one assumes were designed in accordance with current building by-laws, were built from the local chalk. This was used in shuttered form to build the main walls, including both fireplaces and main chimney stacks. The fireplaces had inner linings of fired brick and the chimney breasts were supported on six-inch (150mm) deep reinforced concrete lintels; nine-inch (225mm) diameter fired clay drainpipes were used as flue liners and the upper stacks, above roof level, were constructed of fired brick. It seems likely that this form of construction, with perhaps some minor modifications, should satisfy current Building Regulations requirements if all that is required is a basic open fireplace and flue. For something on a much grander scale, the house illustrated in Colour Plate 4 and Fig. 4.4, was constructed with full Building Regulations approval in 2001/2. It has a massive internal circular chimney stack built entirely of piled cob, which forms a major structural element in what is, in effect, a three-storey house.

Building with rammed earth

With rammed earth, the applied required compactive effort is much greater than that achieved in the wet, piled method, necessitating the use of very robust shuttering. The grading of the soils for rammed earth mixes is not dissimilar to those for wet mixes, although with a lower proportion of fines – silt and clay. No fibrous organic material, such as straw, is included. As with mixes for wet methods, there is some latitude in the actual proportions of the constituent materials, depending to an extent upon the particular nature of each. There should be a relatively high sand content, and just enough clay to act as a binder (see psd curves in Fig. 3.1, Chapter 3). A suitable mix will appear damp rather than wet, probably with a moisture content of 8 to 14 per cent, depending upon the particular soil. Too much moisture will impair the compaction process as, of course, water is not compressible. Any large lumps, and stones greater than 20mm diameter need to be broken up or removed, either by sieving or raking and shovelling. The stock of earth mix needs to be protected from rain, as does the top of the wall as work progresses.

Formwork

As indicated previously, the ramming process necessitates very robust shuttering, as the material needs to be rigidly contained during the building operation. At the start, the formwork is fixed on top of the plinth or underpin course. As with wet mixes, this plinth is necessary to help protect the rammed earth from the effects of splash-back or rising damp. Unlike plinths for wet mixes, most early examples of which comprised inner and outer skins of stonework or brickwork with a rubble-filled core, those for rammed earth need a consistent bearing capacity across their width. This is in order to resist the greater stresses imposed during the compaction process so that no settlement within the plinth occurs.

The selection of the material for the shuttering, the length and height of formwork used at any one time, and the means of ensuring that it remains stable, depend to an extent upon the size and complexity of the building. Traditionally, timber planks were used, normally in the order of 30 to 50mm thick, but more recently, sheet materials such as plywood or metal have become more commonplace.

Normally, when plywood is used it would be around 20mm thick, and the bracings would have to be at closer centres than for the thicker timber planks. Not only does the material need to be able to resist the lateral pressure as the earth is compacted, but also the occasional unintended, though almost inevitable, direct impact from the ramming tool. For this reason, timber planks are still worth considering, since they are replaceable individually in the event of damage, whereas it might be necessary to replace a whole sheet of ply. The usual, and more traditional way of proceeding is to construct a lift of rammed earth over the whole of the plinth or underpin. The shuttering is secured over a length of the plinth, matching its width. Within the shuttering, the earth is rammed in a series of 100 to 150mm layers, each being thoroughly compacted, until the full height of the shuttering is reached. When this length is complete the shuttering is dismantled and re-assembled on the next length of plinth, and so on over the whole length of the walling (see Fig. 4.13). When one lift all around the walling has been completed, the formwork is secured over a length of the next level, and the process continues again all around the walling. This process is repeated until the required wall height has been reached.

Commonly, the formwork is in the order of two to three metres long and up to 900mm high. If it is too large, it will be cumbersome and heavy, increasing the time needed for dismantling and re-assembling between lifts. If, however, it is too small, it will need frequent dismantling

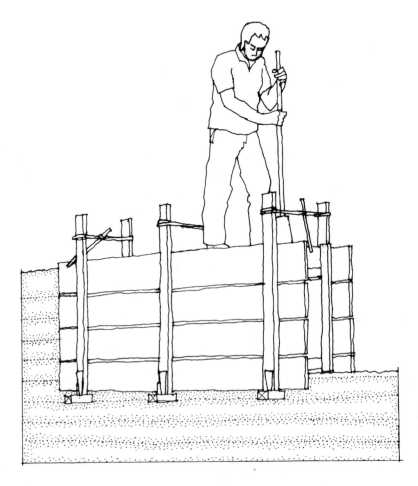

4.13
Traditional method of rammed earth construction

and re-assembly, again increasing the total time needed. Of course, much depends upon the plant, equipment and space available for undertaking the operation. Essentially, the options for keeping the formwork in place as work proceeds are to either make it virtually independent structurally or to secure it with struts or shores transferring the load obliquely to the ground or laterally to existing nearby structural elements such as walls. Of these, the first has always been the most common, as it can be achieved with less construction material and imposes less clutter around the work in progress. It has been estimated that periodic setting up, aligning and then dismantling the formwork is the most time-consuming part of the building process, accounting for around 60 per cent of the total time spent on site.

Generally, using conventional construction techniques, it would be considered impractical to construct shuttering to the full wall height in one lift. Wall widths normally are in the region of 400 to 450mm, although of course this will vary according to height and loading. For such widths, it would

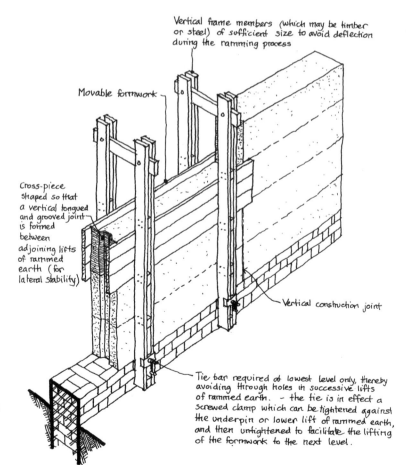

Vertical frame members (which may be timber or steel) of sufficient size to avoid deflection during the ramming process

Movable formwork

Cross-piece shaped so that a vertical tongued and grooved joint is formed between adjoining lifts of rammed earth (for lateral stability)

Vertical construction joint

Tie bar required at lowest level only, thereby avoiding through holes in successive lifts of rammed earth. - the tie is in effect a screwed clamp which can be tightened against the underpin or lower lift of rammed earth, and then untightened to facilitate the lifting of the formwork to the next level.

4.14

Rammed earth; full-height, sliding formwork for continuous construction of vertical panels

be difficult for operatives to undertake the work if the shuttering exceeds 900mm in height. However, techniques developed in Australia and Germany and now being used in Britain would seem to indicate that constructing walls as a series of full-height vertical panels may represent a more satisfactory solution. Although specialised, more robust and therefore more expensive formwork needs to be employed, the result is a more homogeneous wall that takes less time to construct and is less prone to the effects of shrinkage. In order to prevent lateral movement, a vertical slot or shear key may be incorporated in the end stops of the shuttering (see Fig. 4.14). A control or movement joint can, if required, be achieved by incorporating a flexible material similar to that used in the expansion joints of 'conventional' masonry walls. In what is known as the 'Hit and Miss' method, vertical panels are built in two stages. In the first stage alternate panels are constructed to full height along the complete length of the wall. When this stage is complete, earth is rammed into the gaps, thus forming a continuous wall. The effect of this is to reduce shrinkage cracking and the need for end stops in the formwork at the second stage (University of Bath, 2002–4). Productivity rates vary, from one up to three cubic metres per day for a three- to five-man team, depending upon degree of mechanisation, and on-site arrangements for mixing and delivery of the material. Because the 'green' compressive strength of soil, when pneumatically rammed at or slightly above optimum moisture content (OMC) is in the region of $500kN/m^2$, it should, in theory, be possible to construct a free-standing, self-supporting wall in excess of six metres high in one continuous lift. However, in practical terms, a height of two metres is likely to be the maximum height that could be achieved in one day. As for manually rammed walls, the length of individual panels would be between two and three metres. Use of this building system can, however, impose certain design constraints; the width of the vertical panels, which in effect form the basic design module, determining to a certain extent the overall form of the building, the positioning of door and window openings, as well as other structural elements.

It must be emphasised that the traditional types of formwork illustrated in Figs 4.15 and 4.16 are based on systems which have evolved over a period of around 200 years. For small-scale building projects by self-builders, or low-budget community buildings constructed using volunteer labour, such formwork is probably quite suitable. However, for larger projects, such as speculative housing developments and public buildings built under contract, much more sophisticated, heavy-duty formwork systems are now commonly employed, some details of which may be found in various publications included in the Bibliography, in particular Easton (1996), Keable (1996), Minke (2000), Standards Australia (2002) and Walker et al. (2005). So, rather than illustrating and describing all the different methods here, we include an

4.15

Typical arrangement of formwork for manual rammed earth construction

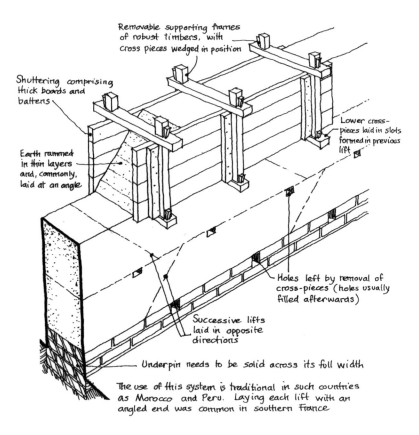

Removable supporting frames of robust timbers, with cross pieces wedged in position

Shuttering comprising thick boards and battens

Earth rammed in thin layers and, commonly, laid at an angle

Lower cross-pieces laid in slots formed in previous lift

Holes left by removal of cross-pieces (holes usually filled afterwards)

Successive lifts laid in opposite directions

Underpin needs to be solid across its full width

The use of this system is traditional in such countries as Morocco and Peru. Laying each lift with an angled end was common in southern France

account of a typical traditional arrangement, identifying the basic elements to achieve stability while enabling ready dismantling and re-assembly of the formwork (see Fig. 4.15). It will be seen that the side shuttering is reinforced with battens. In the case of 30mm timber planking, these battens need to be at about 600mm centres, but probably about 400 to 450mm centres for 20mm plywood. The formwork needs to overlap the plinth or lower lift of rammed earth by between 100 and 200mm. To keep the side shuttering in place it is contained within a removable and adjustable frame. This comprises top and bottom horizontal ties and vertical posts, usually of timber, steel tubing or threaded rod. The posts extend through the ties, and are secured by wedges. At the higher level, sometimes rope is used. This is looped over the upper length of the posts, and can be tensioned by twisting with an inserted rod, as shown in Fig. 4.16A.

In order to ensure that the side-shuttering members remain upright and at the right distance apart as the work progresses, spacing sticks are wedged between them. The formwork to be located at wall ends and returns requires a solid cross-piece or end stop in order to shutter the earth at these locations. The formwork for the wall returns should be a separate unit, since it requires a modified arrangement at its base in order to saddle over the return of the plinth or lower lift of earth walling (Fig. 4.16B). It would normally incorporate a vertical triangular section of timber in order to form a chamfer, which allows more thorough compaction to take place and reduces the risk of erosion or impact damage to the external corners, which are, of course, rather vulnerable.

Sometimes, a vertical shuttering face is included to contain the earth that will abut the next length. An alternative is to compact the earth at an angle, commonly about 45 degrees, so that the next length overlaps it (see Fig. 4.15). If timber planking is used, its inner face needs to be planed so that the earth does not stick to it. In all cases, the inner face should be coated in a release agent so that the formwork can be removed (by sliding it up and to the side rather than pulling straight off), without damaging the face of the earth wall. Slots need to be left across the width of the plinth and finished walling below the current lift in order to accommodate the lower horizontal ties. This is perhaps the one disadvantage of using formwork not braced outside the area of the wall, since it means that material has to be packed into the slots after removal of the formwork.

In south-east France (regions of Rhône-Alpes and Massif Central) where many thousands of such *pisé* buildings exist – dating mainly from the late eighteenth and nineteenth centuries – the soils, which are often tills of glacio-fluvial origin, contain sands, gravels and stones, with a clay content varying from 8 to 16 per cent. Experience having shown that, in the higher clay soils, the horizontal and vertical or diagonal joints between *'banches'* or lifts of rammed earth tended to open up as the walls dried out, builders adopted the expedient of placing a bed of lime/sand mortar between each lift and vertical or diagonal joint as building proceeded. Lime/sand mortar was also used to back-fill the slots left when the formwork was removed. The corners of *pisé* buildings, where abrasion and erosion was most likely to occur, were reinforced with brick, stone or lime/sand mortar, and door and window openings were formed using fired bricks or timber framing.

Methods of compaction

Traditionally, the compaction of rammed earth was achieved through dynamic compression using wholly manual methods. These involved operatives creating an impact on the horizontal surface of the material using purpose-made rammers or tamps. The impact of the rammer produces a shock wave and pressure which sets the particles in motion. The consequence of the ramming is to consolidate the material below the surface and lock it together. In order to ensure even and sufficient compaction, it was normal to have two operatives working over the same area at the same time. They would face one another and ram alternately, with their rammers inclined at an angle. They

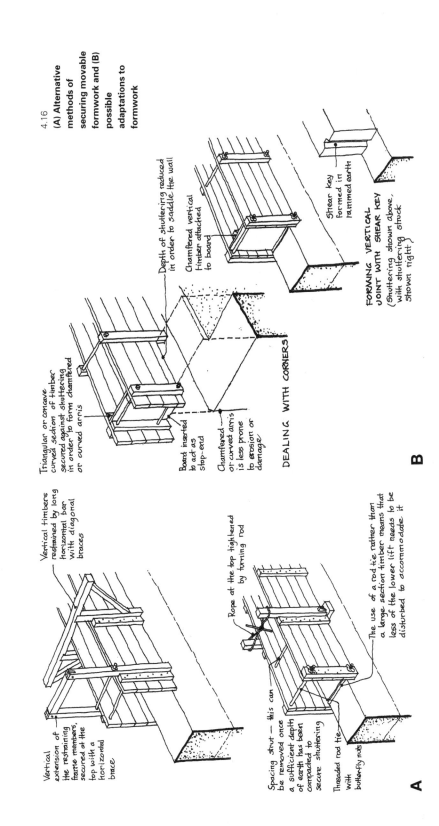

4.16
(A) Alternative methods of securing movable formwork and (B) possible adaptations to formwork

A

Vertical extension of the restraining frame members, secured at the top with a horizontal brace

Vertical timbers restrained by long horizontal bar with diagonal braces

Rope at the top tightened by turning rod

Spacing strut — this can be removed once a sufficient depth of earth has been compacted to secure shuttering

Threaded rod tie with butterfly nuts

The use of a rod tie rather than a large section timber means that less of the lower lift needs to be disturbed to accommodate it

B

Triangular or concave curved section of timber secured against shuttering in order to form chamfered or curved arris

Depth of shuttering reduced in order to saddle the wall

Chamfered vertical timber attached to board

Board inserted to act as stop-end

Chamfered or curved arris is less prone to erosion or damage

DEALING WITH CORNERS

Shear key formed in rammed earth

FORMING VERTICAL JOINT WITH SHEAR KEY
(Shuttering shown above, with shuttering struck shown right)

would also ram at a slight angle towards the shuttering, so that the material became adequately and evenly compact throughout the whole width of the wall. Although, traditionally, hardwood ramming tools were employed, nowadays it is more usual to use tampers with steel ramming heads. In Fig. 4.17 the tampers shown at A and B are conventional all-purpose tools, whereas the one shown in Fig. 4.17C, which may have either a steel or hardwood ramming head, is designed specifically for compacting the soil at the sides and corners of the formwork. The average weight of a hand tamper is about 5kg and its overall length 1.5 metres.

The manual process is particularly tiring and time-consuming, and would no longer be seen as an appropriate and economic building operation except perhaps in self-build projects or in those places where labour-intensive building activities remain the norm. It is for this reason that, for major building projects, mechanised ramming tools are now normally employed. Probably the most satisfactory of these has been the pneumatic rammer. Pneumatic rammers are available from several manufacturers, the best known of which are Ingersoll-Rand and Atlas-Copco. They are, of course, much heavier than hand tampers, around 11kg. They have a compaction stroke of about 200mm and operate at a rate of about 700 cycles per minute,

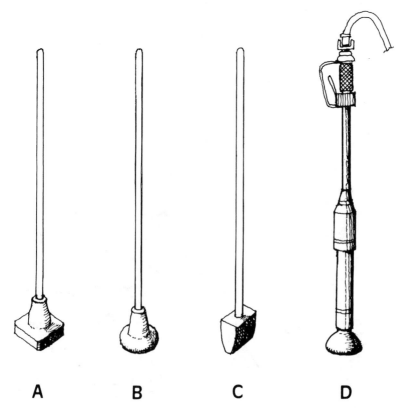

A   **B**   **C**   **D**

4.17
Tampers for rammed earth construction

allowing work to proceed at around four times the rate achieved by manual ramming, therefore saving significantly on labour costs. However, operating these machines for long periods of time is physically quite demanding. Potentially harmful effects may include temporary deafness, 'vibration white finger' and various repetitive-strain injuries, although these can be controlled to a great extent by good site practice and the use of protective clothing. A typical pneumatic rammer is shown in Fig. 4.17D.

It has been claimed that an alternative, more effective way of achieving dynamic compression is through vibration. The use of plate vibrators is long established in the civil engineering industry, particularly in roadworks. By producing a series of rapid impacts on the soil at a rate of 1,000 to 1,200 vibrations per minute, pressure waves are generated which extend well beneath the surface of the material. The pressure waves impart motion to the particles, temporarily overcoming internal friction, and resulting in a rearrangement of the grains in a much denser formation where they are adequately constrained laterally. In some situations the vibration technique would enable the earth to be evenly compacted to a greater depth than can be achieved through hand or pneumatic ramming. Conventional plate vibrators were designed specifically to compact and consolidate the dry or semi-dry non-cohesive materials used as a sub-base in highway and pavement construction and are not generally considered suitable for the compaction of clayey sands. However, specially designed electrically driven vibrators have been developed and successfully used for rammed earth construction in Germany (Minke, 2000).

Chalk, dug on site, was used for rammed earth dwellings in an experimental housing development in Amesbury, Wiltshire. These were constructed in 1919–20 in a project undertaken by the Department of Scientific and Industrial Research for the Ministry of Agriculture. In all eight of the chalk-based houses (including a semi-detached pair) built at Amesbury, the chalk was used in a modified form. One was built using the traditional 'wet, piled' method, with straw and water being added to the chalk, while another was built using a similar but drier mix rammed into timber shuttering.

Four houses, including the semi-detached pair, were built of what was called 'chalk *pisé*' – a mixture of 70 per cent chalk and 30 per cent soil, but containing no straw, also rammed into timber shuttering. Of the two remaining houses, one was built of fibreless, rammed chalk stabilised with 5 per cent Portland cement, and the other had cavity walls constructed from pre-formed blocks comprising 12 parts chalk to one part of Portland cement.

Because of their greater compaction and density, rammed earth walls are generally stronger than those built using wet methods. Also, the higher density makes the material more resistant to subsequent water absorption, thereby reducing the risk of excessive moisture movement and softening. However, it has to be recognised that, in rammed earth construction, more material is used than for a wet method earth wall of the same volume. Also, given that a similar grading and quality of materials is used, the higher density would result in a higher thermal conductivity. Further discussion of these comparisons is included in Chapter 5.

In recent years, the ramming technique has proved popular for new earth-walled buildings in Britain. Probably the main reasons for this are its speed of construction, greater compressive strength enabling thinner walls, reduced drying shrinkage, and ease of exercising quality control when compared with wet construction techniques. Recent examples include a number of dwellings, a building at the Centre for Alternative Technology in mid-Wales, and the visitor centre at the Eden Project in Cornwall (see Colour Plates 7 and 8).

Building with earth blocks and clay lump

Unfired earth bricks (adobes) have never formed part of the mainstream British vernacular building tradition. The extensive use of clay lump in parts of East Anglia represents, in historical terms, a relatively recent innovation, their use having first been recorded during the last decade of the eighteenth century. Despite the widespread use of unfired earth bricks and blocks throughout the world, especially in developing countries, the authors are aware of only one recent public building project in Britain where this construction method has been employed, and this is a toilet block at the Centre for Alternative Technology in mid-Wales, built in 1999.

Adobes and compressed soil blocks can be used in any situation where fired bricks are normally employed, and most skilled, experienced bricklayers or masons can easily adapt to using these 'unconventional' materials (see Fig. 4.18). In the hot, arid regions of the world, where there is little rainfall and timber for building is in very short supply, sun-dried mud bricks have been used to construct domed and vaulted roofs – a technique which has its origins in the Middle East, North Africa and, most particularly, in Egypt. These spectacularly complex roofs, which are the result of thousands of years of development and refinement are, unfortunately, not entirely suited to the cold, damp conditions that prevail in many parts of North America and northern Europe, and are, therefore, outside the scope of this book. For those who wish to find out more about these fascinating construction techniques there is a wealth of published material available and the technique is fully described in Minke (2000).

The most appropriate mortar for use with unfired earth bricks and blocks is made simply from sieved earth, mixed with sufficient water to

4.18
Building earthquake-resistant housing in Yemen using compressed soil blocks
Photo: David Webb

achieve the required consistency. In order to maintain a degree of homogeneity, the same soil as that used to fabricate the blocks should be
employed to make the mortar. As well as having very similar physical characteristics to the bricks and blocks they are intended to support, earth mortars
have much greater cohesion to earth substrates than do the more commonly
used lime/sand mortars. The only case where the use of anything other than
earth mortar might be justified is when blocks have to be laid in cold, damp
conditions; in which case a weak hydraulic lime, say 20 per cent NHL-2
(European standard for natural hydraulic lime with a minimum crushing
strength of 2N/mm²), together with an equal amount of dry, sharp sand might
be added to the soil (proportions of 1:1:3) in order to achieve a set. As with
earth plasters, which are discussed in the following section, clay-rich soils
can pose problems, for two reasons. Firstly, they may shrink to an unacceptable degree, leaving cracked, partly open joints and, secondly, they may
be so sticky as to be virtually unworkable, and very difficult to remove from
the trowel. The workability of such soils can be improved by the simple expedient of adding a sufficient quantity of sand.

Adobe bricks, clay lumps and, to a lesser extent, cob blocks,
because they are formed from wet material, will tend to be rather variable in
shape, and in order to allow for this they need to be laid in a fairly thick (20
to 25mm) mortar bed. Compressed soil blocks, on the other hand, are formed

at a low moisture content to precise dimensions and may, with care, be laid in a very shallow mortar bed, almost like a slurry or even, if well wetted, laid with no mortar at all. It is important, when building with any type of earth block, to ensure that all contact surfaces are kept in a damp condition so as to reduce suction, thereby eliminating the risk of too rapid drying out and partial loss of adhesion through the development of shrinkage cracks. As previously noted, earth blocks can be used in conjunction with mass earth walls and in this context have been widely used for structural repair works. They are useful for constructing fairly slender (down to 150mm) structural and non-load-bearing internal walls as they have a high resistance to the transmission of air-borne sound. Lightweight earth blocks, having a dry density of < 1,100kg/m^3, can be used to form an external skin to mass earth walls in order to improve their thermal resistance or 'R-value' (see Chapter 5).

Earth-based finishes for earth walls

The use of lime/sand mortars for plastering and rendering earth walls in Britain is well known and well documented (Holmes and Wingate, 1997; DEBA, 2002) so will not be discussed at great length here. Although common in many parts of the world, earth or 'mud' external renders have never been widely used in Britain, apart from in East Anglia, where the use of earth and earth/lime renders was quite widespread. Earth/hair and, to a lesser extent, earth/chopped hay plasters were applied to the internal walls and ceilings of both cob and random stone buildings from the earliest times up to the end of the nineteenth century. Their widespread use in the medieval and early modern periods could probably be explained by the difficulty and expense involved, especially in remote inland areas, in obtaining building lime and good quality sand.

However, the main reason that earth plasters were so popular and continued in use for several centuries, even after lime and sand became more readily available, was that, for internal use, they were not only much cheaper but provided a finish that was durable, flexible and easily maintained. In south-west England virtually all decorative plasterwork, from the sixteenth up to the mid-nineteenth century, was applied to earth, or 'daub', plasters on riven oak lathing; mainly because the craftsmen, understandably, wanted their work to last and were aware of the proven toughness and flexibility of earth/hair plasters as a substrate and support for decorative lime plasters. It is the hair (normally cow hair) combined with the clay binder that gives the material its considerable tensile and flexural strength. A properly mixed and applied earth/hair plaster is very difficult to pull apart by hand, even when it has been immersed in water for several hours. Furthermore, it is usually quite difficult to separate the layer of applied earth plaster from its substrate, especially if this is a cob wall, because of its very strong adhesion, even when, as is often the case, it has been *in situ* for three to four centuries. A serviceable earth plaster mix for internal use would contain the following proportions (by weight): fines (silt and clay) 40 per cent, fine to coarse sand 40 per cent, grit (2–5mm diameter) 15 to 20 per cent, and either hair or chopped hay 3 to 5 per cent (note that this is equivalent, by volume, to a mixture of soil and organic matter in more or less equal parts).

Tests carried out to measure and compare the adhesion, or bond strength, of earth, lime and cement-based mortars applied to earth walls, have shown that lime/sand mortars fail (pull away from the substrate) at a tensile stress of around $0.09N/mm^2$, while earth mortars fail at between 0.12 and $0.13N/mm^2$. Cement/sand mortars have a very high initial bonding strength, $> 0.16N/mm^2$, but when, subsequently, thermal movement takes place and moisture collects at the wall face, this adhesion can be lost quite rapidly. (Cement-based plasters have poor adhesion to earth walls, which is why they require a mechanical key, usually chicken wire or metal lathing.) A typical internal earth plaster will vary in thickness from 5 to 6mm (early work) up to 10 to 12mm in nineteenth-century buildings. Often it will be finished either with a pure lime or 1:1 lime/fine sand skim coat, 1 to 2mm thick, or with numerous coats of limewash. Earth plasters usually seem to have been applied in one coat. In addition to their cohesion and flexibility, earth plasters, in particular those containing lightweight aggregates such as expanded perlite, will readily absorb moisture from the air and are, therefore, useful in preventing internal condensation and in regulating relative humidity to optimum levels in domestic dwellings.

Composite earth/lime renders

External renders comprised entirely of earth will, to a greater or lesser degree, depending on their particle size distribution (psd) characteristics, be subject to damage as a result of weathering and erosion. They may, therefore, be regarded as 'sacrificial' rather than permanent wall coatings, requiring fairly frequent renewal unless protected in some way. Traditionally, in south-west England, cob walls facing east or north were usually left unprotected (see Fig. 4.19) while those facing south and west, which were exposed to wind-driven rain for much of the year, received a protective coat of limewash. This needed renewing annually and, over a period of many years, could attain a thickness of several millimetres. From the beginning of the nineteenth century, increasing use was made of lime/sand renders for practical, aesthetic and socio-cultural reasons; exposed earth walls being associated with poverty and low status.

In order to improve the performance and durability of earth renders they can be stabilised with non-hydraulic slaked lime, $Ca(OH)_2$, in either

4.19
**North-facing wall of a
mid-nineteenth-century
cob barn near Exeter,
Devon, which has
remained exposed
since being built**

bagged, powder form, when it is known as hydrated lime, or as mature 'lime putty'. The addition of relatively small amounts of lime to well-graded sandy soils with a minimum clay content of around 10 per cent can improve their durability and resistance to erosion in several ways. When lime is added to a clay soil in the presence of water, a chemical reaction will take place between the calcium and the clay minerals. The most immediate and obvious effect, in cohesive clay soils, is that the lime flocculates the clay particles (this process has been described in Chapter 2) so that the soil becomes less plastic and therefore easier to work.

The fine soil and clay particles having become aggregated, the plastic limit increases and the plasticity index is reduced. The soil will take up less water, so that shrinkage may be significantly reduced, by around 60 to 75 per cent; clearly an advantage in the application of plasters and renders. For as long as the soil/lime mix remains damp, calcium will combine chemically with the silica and alumina in the clay minerals to form complex aluminium and calcium silicates, or cementitious compounds. This leads to a significant increase in the strength of the binder fraction, which continues over a 'curing' period of at least seven days, during which time the material must be kept damp. It should be noted that if temperatures are high the gain in strength, provided damp conditions are maintained, will be more rapid. In addition to this process, over a much longer period of time free lime in the soil will combine with carbon dioxide from the air to form calcium carbonate ($CaCO_3$) thus further increasing the soil's resistance to erosion. The amount of lime required varies according to the size of the clay fraction, from 10 to 20 per cent, by volume, of the dry soil. For well-graded clayey sand soils, around 10 per cent (1:9 lime/soil) would be considered appropriate, but for soils containing a large clay fraction, in excess of 20 per cent, a lime/soil proportion of 1:5 is recommended. Soils containing large amounts of silt should be avoided, mainly because silts are chemically non-reactive and serve only to fill up the soil voids, thus reducing permeability.

Mixing and application

As with lime-based plasters, it is always best to do two or three experimental mixes, containing variable amounts of water and, if necessary, added sand and apply these to 500mm square trial panels so that drying, shrinkage and adhesion can be monitored and measured. Local subsoil or reconstituted cob is first dried and pulverised, then passed through a garden or mason's sieve with a mesh aperture of 5 to 6mm. Subsoils containing more than about 40 to 50 per cent of fines (clay and silt) may need to be gauged with coarse to medium sand, as otherwise they may be too sticky to be workable and may develop shrinkage cracks too wide to be controlled by subsequent reworking. Earth plasters and mortars may be mixed manually on a hard, smooth surface – a sheet of exterior-grade plywood, for example, or, if large quantities are required, a pan mixer may be employed. It is important to ensure that the large quantity of hair required should be incorporated in such a way that it is evenly distributed throughout the mix.

Earth plasters, like lime plasters, should be applied to a thoroughly pre-wetted surface. In cob walls having hollows and cavities, these should first be 'dubbed out' with a subsoil/chopped hay or straw mix, keyed into the cob. Earth mortars are very cohesive and, in order to be workable, may need to be fairly wet, around 20 per cent or more water content. Unless gauged with sand, clay-rich plasters may be almost impossible to apply. (Plasters containing chopped hay are easier to apply than loam/hair plasters, but are probably less durable.) Earth renders are used in the renovation of cob buildings in Normandy, in north-west France. They are usually gauged with 10 per cent lime and are mixed and applied mechanically, being projected on to the external walls using a compressed air hose, then floated by hand in order to control shrinkage. In a variation of this technique, developed in Arizona, USA, two-inch (50mm) thick layers of sieved earth are blown through a hose on to adobe and rammed earth wall surfaces using a machine described as a 'mud pump'.

Although throwing (or casting) the material on to the wall surface, either mechanically or by hand, is known to improve adhesion, earth plasters and renders are most usually applied with a steel trowel or float, using a normal lime plastering technique. If timber or fibreglass floats are employed, they will need to be kept well wetted. On internal walls, surface cracking is usually less of a problem than it is with externally applied renders. Repeated re-working of the material should be avoided, as this will bring the coarse aggregates and hair to the surface. Fine surface cracks on internal walls can be filled either with limewash or an applied lime-based skim coat (1:1 lime putty/fine sand < 1.0mm diameter).

The addition of small quantities of lime to soil requires careful and thorough mixing to ensure that the lime is well distributed throughout the soil. Adding more than the recommended amount, rather than increasing strength will tend to have the opposite effect and should, therefore, be avoided. Either hydrated lime, in bagged, powder form, or lime putty may be used, though adding hydrated lime to dry soil prior to wet mixing might enable the mix proportions to be more accurately gauged. The mixing procedure is much the same as that for earth plasters and renders, but with one important difference: the material should not be applied to the wall straight away but be left under cover for two or more days in order that the clay/lime reactions described above can take place and then, if necessary, be re-mixed before being applied to the wall.

Paint finishes

When a paint finish is applied to any form of breathable raw earth wall it should, ideally, have water vapour permeability characteristics similar to those of the earth or earth/lime substrate that it is intended to protect. Traditionally, limewash has always been used to provide a protective but sacrificial coating to earth walls because of its high permeability and the way in which it reacts with the clay particles in the soil to provide a degree of consolidation at the wall surface. The mixing and application of limewash is not discussed here because there is a wealth of detailed information available in other publications (Schofield, 1994; Holmes and Wingate, 1997). The only other paint finish with permeability characteristics similar to those of limewash is a potassium silicate-based product known as Keim Granital, manufactured in Germany by Keimfarben GmbH. It is very durable, lasting 15 to 30 years before needing replacement, but also rather expensive and has to be applied strictly according to the manufacturer's instructions. It is used primarily as a two-coat application for painting rendered masonry walls and, to the authors' knowledge, there are no examples of it having been applied directly to unrendered earth walls.

1.
Sculptural: Part of the Great Mosque at Djenné, Mali, built 1906–7. The building is 75 metres (250 feet) square
(photo: Mike Smallcombe)

2.
Monumental: The fortified earth city of Shibam in central Yemen, parts of which date from the sixteenth century
(photo: Howard Meadowcroft)

3.
Sculptural: Public toilets and bus shelter at the Eden Project, St. Austell, Cornwall 2002.
Designers/builders Abey/Smallcombe and Back to Earth (photo: Mike Smallcombe)

4.
Organic: New cob house at Ottery St. Mary, Devon 2000–1.
Designer/builder Kevin McCabe (photo: Kevin McCabe)

5.
Neo-vernacular: Two storey cob extension to seventeenth century cottages at Down St. Mary, mid-Devon 1990–1.
Designer/builder Alfred Howard

6.
Contemporary cob: Information centre and public toilets at Norden park-and-ride facility, Corfe Castle, Dorset 1998.
Architect
Robert Nother

7.
Rammed earth wall at the Eden Project visitor centre, Cornwall 1999. Ninety metres (295 feet) long and made up of vertical panels 2.4 m high, built up off a concrete foundation/plinth.

Architects Nicholas Grimshaw Partnership; Contractor In Situ Rammed Earth Company (photo: Peter Walker)

8.
Internal load-bearing rammed earth wall, 4 metres high, at the Centre for Alternative Technology, Machynlleth, Wales 1999.

Architect Pat Borer; Contractor Simmonds-Mills (photo: Peter Walker)

Chapter 5

Standards and regulations

By R. Nother, with a note on Standards and Codes of Practice by L. Keefe

Introduction

In this chapter we shall be looking at the potential for using unbaked load-bearing earth in new-build situations, whether for stand-alone buildings and structures or extensions to existing earth-walled buildings. As indicated previously, there is a common perception amongst practitioners, certainly in Britain, that statutory standards limit the potential for using such materials. Compared with some European countries, Britain probably is lagging behind a little in their use, and lacks laid down standards specifically against which to assess its qualities. However, there are now several research institutes and associations aiding and encouraging the use of unbaked earth, some of which are listed at the end of this book under Organisations. Britain appears to be catching up, and with the gradual unification of standards across the European Community, one can expect an increasing willingness to use the material. To greater or lesser degrees, virtually all countries have statutory controls over the construction and location of buildings. In this chapter, the focus is on those applicable in England and Wales, since those in many other countries have essentially similar or less stringent provisions. In Britain, the main statutory controls relating to new buildings are the Town and Country Planning Acts and the Building Regulations. The application of the Building Regulations to earth-walled buildings is examined later in the chapter, but firstly we look at the implications of trying to satisfy planning legislation.

Planning legislation, policies and guidance

In England and Wales, the principal planning legislation is the Town and Country Planning Act 1990. The Planning (Listed Buildings and Conservation Areas) Act 1990 also applies. Supporting these Acts are numerous Regulations and Orders. Largely, guidance on the application of the Acts can be found in a series of Planning Policy Guidance notes issued by the Department of the Environment, Transport and the Regions (or its predecessors).

Planning Policy Guidance Note 1 (PPG1, revised February 1997) is entitled 'General Policy and Principles', and explains that the PPGs set out the Government's policies on different aspects of planning. It states that 'Local planning authorities must take their content into account in preparing their development plans. The guidance may also be material to decisions on individual planning applications and appeals.' Amongst other things, PPG1 emphasises the contribution of the planning system to achieving sustainable development, considering this to be one of its key roles. In PPG1 the Government states its commitment to the principles of sustainable development set out in *Sustainable Development: The UK Strategy* (1994). The *Strategy* recognises the important role of the planning system in regulating the development and use of land in the public interest. The formulation of a sustainable planning framework involves a complex range of considerations, one of which is how new development can be shaped in a way which minimises the need to use environmentally costly transportation. Largely, this may be considered to mean enabling people to live close to their places of work, but also important is the need to ensure that, where practical, materials for building construction are not transported over great distances. This connects with another section of PPG1 dealing with design, in which the worthiness of seeking to promote or reinforce local distinctiveness is recognised. Traditionally, this has been achieved through using building materials, details and forms local to a particular area. As has been shown previously, earth walling survives in significant amounts in many parts of the UK, and once was much more widespread. It has been shown also that earth walling scores highly in sustainability terms in its energy-efficient production.

There are other PPGs which stress the need for development to be sustainable. For example, in PPG3: Housing (March 2000), this forms a major part of the document. Whilst the emphasis appears to be on land use, density and location of housing development, the choice of construction materials can also be seen to be a factor; for example, where endeavouring to achieve sympathetic design in sensitive locations. PPG12: Development Plans and Regional Planning Guidance (February 1992) contains a chapter devoted to sustainable development. This includes reference to the document *A Better Quality of Life, a Strategy for Sustainable Development in the*

UK (1999), and emphasises that the planning system, and development plans in particular, can make a major contribution to the achievement of the Government's objectives for sustainable development. Of course, the emphasis in development plan policies should be on the land-use aspects of sustainable development. However, a holistic approach is called for, and there are important environmental considerations which could place a focus on the choice of building materials. For instance, PPG12 indicates the need to sustain the character and diversity of the countryside, and to consider the impact of development on landscape quality. In this, the choice of external materials and the manner of their use may be key factors in assessing whether a proposed development is acceptable.

Another particularly relevant document is PPG15: Planning and the Historic Environment. The introductory content repeats the Government's commitment to the concept of sustainable development, and notes the particular relevance of this commitment to the preservation of the historic environment. While to a considerable extent this may be concerned about the need to recognise the value of surviving historic fabric, it is also about built places and the need to assess the impact of proposals for new development on the historic environment. Often, places gain their special local character in part through the use of local building materials. In many places, earth walling is a key ingredient of the local palette, and implicit in the guidance contained in PPG15 is the encouragement of its continued use in new development.

In terms of planning considerations, then, there is no reason why earth-walled buildings should not be included in many areas where new development is deemed acceptable. Indeed, there may well be situations where earth walling and the building forms associated with it give an opportunity to arrive at more comfortably contextual solutions than may be achieved with other walling materials. Some earth-walling techniques engender a certain plasticity of form and solutions having an organic appearance. This is very much the case with the traditional cob and thatched dwelling, either singly or in groupings. It may be that, rather than replicate these early examples, there is an opportunity to develop contemporary solutions that sit comfortably alongside their established neighbours.

Building Regulations

The planning system is not generally concerned with the actual construction techniques of new buildings except in so far as they influence their external appearance. The physical attributes of new buildings are more the concern of the Building Regulations. In essence, the Building Regulations are concerned with the construction in terms of its physical qualities and energy consumption, ready access for people with disabilities, the safety of those who use the building or who are near it, and the impact that it will have upon certain physical qualities of the environment.

The Building Regulations for England and Wales applicable at the time of writing are The Building Regulations 2000 and The Building (Amendment) Regulations 2001. From time to time the Building Regulations are revised: for example, over recent years, thermal insulation requirements have been increased, and are likely to be further increased in the future. The Building Regulations are supported by a set of Approved Documents. Each of these repeats the particular section of the Building Regulations to which it refers, and then provides guidance on how compliance with the Regulations might be achieved. It must be stressed that the Approved Documents are simply guidance, and that there may be other ways of achieving compliance with the Regulations.

The Building Regulation Approved Documents

The titles of the Approved Documents correspond with the titles set out at Schedule 1 of the Building Regulations, and are as follows:

A – Structure
B – Fire safety
C – Site preparation and resistance to moisture
D – Toxic substances
E – Resistance to the passage of sound
F – Ventilation
G – Hygiene
H – Drainage and waste disposal
J – Combustion appliances and fuel storage systems
K – Protection from falling, collision and impact
L1 – Conservation of fuel and power in dwellings
L2 – Conservation of fuel and power in buildings other than dwellings
M – Access and facilities for disabled people
N – Glazing – safety in relation to impact, opening and cleaning
Approved Document to support Regulation 7 – Materials and Workmanship

Part P, which is concerned with electrical safety in dwellings, came into effect on 1 January 2005. Also under consideration is the proposal for a Part Q, which will deal with broadband services. This is in recognition of the increasing use of electronic communication services, and that it is more practical to provide broadband service cables in new buildings rather than to install them retrospectively.

Additional supporting documents are 'Thermal Insulation: Avoiding Risks', and 'Support Document Part 1, 2001, Robust Details' (the latter showing examples of construction which are offered as a means of achieving the required levels of thermal insulation while avoiding the risks of surface and interstitial condensation). The main Approved Documents applicable to the choice of walling material are A, B, C, E, L1 and L2. Approved Document G, which covers hygiene, does not apply to the actual building fabric; instead, it is concerned with sanitary conveniences and washing facilities, bathrooms and hot water storage. Of key relevance are the Approved Document to support Regulation 7 – Materials and Workmanship, and the supporting documents dealing with thermal insulation and condensation.

The following content examines these documents and the impact that they might have upon the acceptability or otherwise of earth-walling types. Firstly, let us examine Regulation 7, which covers materials and workmanship, and states:

> Building work shall be carried out –
> (a) with adequate and proper materials which –
> are appropriate for the circumstances in which they are used; are adequately mixed and prepared; and which are applied, used or fixed so as adequately to perform the functions for which they are designed; and
> (b) in a workmanlike manner.

Approved Document to support Regulation 7 – materials and workmanship

The guidance in this Approved Document starts with introductory guidance on performance, followed by Sections 1 and 2 dealing respectively with materials and workmanship. In terms of performance, environmental impact of building work is seen as an important consideration. The Approved Document guidance advises that:

> The environmental impact of building work can be minimised by careful choice of materials, and where appropriate the use of recycled and recyclable materials should be considered. The use of such materials must not have any adverse implications for the health and safety standards of the building work.

In defining the range covered, the guidance includes reference to naturally occurring materials. Thus, unbaked earth-walling materials can be deemed to be included within the acceptable range provided that they meet the other required criteria. This would embrace the need to ensure that the material is safe and hygienic, and there is no reason why earth-walling materials cannot be made to satisfy such criteria. Largely, this will involve taking a careful approach in the selection and checking of the constituent materials for the load-bearing element and the applied finishes. In all probability, the perception that earth-walling is unhygienic stems from the days of the 'improvers' of the nineteenth century, when many of the smaller earth-walled dwellings did not receive an appropriate level of maintenance, and acquired a substandard appearance.

Most building materials used these days are either factory produced or are natural materials that have been subjected to approved tests prior to site delivery. Much of the Approved Document guidance on the compliance of building materials with Regulation 7 is concerned with the recognised standards for testing and displaying conformity. There is coverage of British Standards, other national and international technical specifications, technical approvals, CE marking, and independent certification schemes. Additionally, materials are acceptable which can be shown by tests, calculation or other means to perform their designed function, provided that tests are carried out by independent accredited testing laboratories.

Also included is recognition of the acceptability of past experience. A material is acceptable if it can be shown by experience, such as its performance over time in a building in use, to be capable of performing the function for which it is intended. Broadly, the content of Section 2 dealing with Workmanship covers provisions similar to those for materials. Additionally, for workmanship there is a paragraph covering Management systems. This includes reference to BS EN ISO 9000; Quality Management and Quality Assurance Standards. Also recognised are some independent schemes for accreditation and registration of installers of materials, products and services that provide a means of ensuring that work has been carried out by knowledgeable contractors to appropriate standards. Hence, while it may appear to be a little less straightforward to achieve conformity to Regulation 7 for earth-walling materials than other more commonly used materials, they are certainly not excluded. Much hangs on the development over the next few years of accredited standards agencies, and considerable work is being undertaken at present to achieve this. However, in the absence of approved British Standards, Section 1 of Regulation 7 does allow the submission, in support of applications for Building Regulations approval, of technical specifications and approvals from other member countries of the European Union, which, in the case of earth construction, may be quite helpful. The issue of technical standards and codes of practice for earth building are further discussed at the end of this chapter.

Approved Document A – structure

Schedule 1 of the Building Regulations sets out the requirements under Regulations 4 and 6, and Part A of this schedule is concerned with structure. Part A1 deals with loading, and requires a building to be constructed so that the combined dead, imposed and wind loads are sustained and transmitted by it to the ground safely and without causing such deflection or deformation of any part of the building, or such movement of the ground, as will impair the stability of any part of another building. In meeting these requirements, account has to be taken of the imposed and wind loads to which the building is likely to be subjected in the ordinary course of its use for the purpose for which it is intended. This is in addition to the self-weight of the structure. Part A2 deals with ground movement, essentially requiring the building to have foundations designed to counteract any ground movement which otherwise might impair the stability of the building. There is also a Part A3 which deals with disproportionate collapse, this being applicable only to buildings having five or more storeys. For the purposes of this chapter, it is assumed that generally the earth-walled buildings being considered are of relatively low-rise, and therefore that Part A3 will not be applicable. The authors are unaware of any existing earth-walled buildings in the United Kingdom which have five or more storeys, although in other parts of the world, notably the Yemen and North Africa, there are historic earth-walled buildings which exceed this number of storeys (see Colour Plate 2).

Part A2, dealing with ground movement, may be considered to be relatively straightforward in terms of achieving compliance. At Section 1E of Approved Document A, a number of conditions and design provisions are set out relating to strip foundations of plain concrete, compliance with which would satisfy the Building Regulations. Table 12 of the Approved Document shows the minimum width of strip foundations for various types of subsoil and a range of total loads for load-bearing walls. It is assumed here that concrete would be the usual foundation choice for contemporary load-bearing earth walls, associated with some other substructure material such as brick or block which enables the earth walling to be kept above ground level. Because earth walls generally are thicker than the equivalent conventional cavity brickwork or blockwork wall, correspondingly the concrete strip foundation also will be wider. In most cases, therefore, the strip foundation is likely to be of more than adequate width to transfer the building load to the ground. In traditional earth buildings, as with other forms of traditional load-bearing wall construction before the advent of mass concrete, the load was transferred to the ground through a spreading of the substructure material over a wider area at the base. There is no reason why these more traditional substructure arrangements could not be used today provided that they can be shown to transfer the building load to a bearing strata capable of carrying it and at a depth to eliminate the risk of movement. However, this would involve submission of structural calculations relating to the detailed design proposals, as it would not conform to the approved document examples.

A more frequent concern is likely to revolve around the cost of providing a greater than normal width of substructure and plinth walling to carry the cob-type wall. However, this is not an issue relevant directly to the Building Regulations, and ways of achieving economy in substructure design will be looked at later. The main point to stress here is that foundations for earth walls are as readily achievable for compliance with the Building Regulations as they are for other load-bearing wall types. Where the ground conditions are such that the guidance in Approved Document A2 cannot be applied, it will be necessary to consider alternative foundation structures, just as would be the case with other walling types. In such instances, the services of a structural engineer are likely to be required, but the design is unlikely to have to be any more onerous than for other load-bearing walling types.

As an indication of how the requirements of Part A1 might be met, Approved Document A describes and illustrates a number of conditions. Largely, these deal with slenderness ratios (i.e. structurally effective thickness of wall in relation to its structurally effective height), lateral support, buttressing, and number and sizes of openings. Generally, there should be no great difficulty in achieving conformity with these conditions, since usually earth walls such as cob are considerably thicker than their brickwork or blockwork counterparts and can be readily designed to satisfy slenderness ratio requirements in low-rise, domestic-scale construction. When using shuttered materials, such as rammed earth, consideration needs to be given to the cost and speed of erecting the wall in assessing the choice between a thinner wall with buttresses and a thicker wall of constant width. The introduction of buttresses is likely to add considerably to the complexity of the shuttering, thereby increasing the cost and slowing down the construction process. This needs to be balanced against the cost of the additional material that may be required for a wall of continuous width. Of course, there are likely to be other factors besides structural requirements which influence the choice of thickness, such as the level of sound or thermal insulation required.

In terms of the actual load-bearing capacity, the conforming examples of external walls shown in the Approved Document are confined to cavity brickwork and blockwork, and therefore are not likely to apply to structural materials such as cob and rammed earth. It is likely, therefore, that the structural adequacy of the earth wall will have to be shown by calculation. This will involve making due allowance for live and superimposed loads, and for the self-weight of all the constructional elements, including the wall itself. The greatest compressive stresses in the walling will occur at bearing points of floors and lintels, piers between openings, and at its base. To determine the self-weight of the wall, its density will need to be known, and this can be obtained from weighing a matching sample. The density of a mass earth wall will vary from 1,600 up to 2000kg/m^3, according to soil type and method of construction. The load-bearing capacity of a 550mm cob wall will vary according to its density, but on average may be taken as 750kN/m^2. A two storey residential building will have an average stress in the wall of 160kN/m^2. The loading at the base of the wall will be approximately 68kN/m, resulting in a ground bearing pressure of 90kN/m^2 with a 750mm wide foundation. These loadings and stresses are more than adequate to meet the requirements for structural safety (Ley and Widgery, 1997). Rammed earth walls can be somewhat narrower for the same load, subject to satisfying slenderness ratio requirements. Clay-lump walls tend to be constructed much narrower than mass earth walls, and should therefore have a compressive strength of at least 2,000 kN/mm^2.

Approved Document B – fire safety

Part B of the Building Regulations deals with fire safety. Part B1 is concerned with means of warning and escape, B2 with internal fire spread (linings), B3 with internal fire spread (structure), B4 with external fire spread, and B5 with access and facilities for the fire service.

Parts B1 and B5 relate largely to spatial and servicing matters, and B2 to surfaces rather than the structural elements. Hence, B3 and B4 are the parts with which we need to be primarily concerned here.

Part B3 requires that a building shall be designed and constructed so that, in the event of fire, its stability will be maintained for a reasonable period. It also requires that a wall common to two or more buildings shall be designed and constructed so that it adequately resists the spread of fire between those buildings. There are also requirements to subdivide a building with fire-resisting construction to an extent appropriate to the size and intended use of the building, and to design and construct the building so that the unseen spread of fire and smoke within concealed spaces in its structure and fabric is inhibited. The requirements of B4 are that, firstly, the external walls of the building shall adequately resist the spread of fire over the walls and from one building to another, having regard to the height, use and position of the building. Secondly, the roof of the building shall adequately resist the spread of fire over the roof and from one building to another, having regard to the use and position of the building.

To satisfy the requirements of Part B3, there are four conditions which need to be met. These deal with the load-bearing elements of the structure, dividing the building into fire-resisting compartments, protecting openings in fire-separating elements, and sealing and subdividing any hidden voids in the construction to inhibit the unseen spread of fire and products of combustion. Our main concern here is with the first of these, which covers structural load-bearing elements such as structural frames, floors and load-bearing walls. It needs to be shown that the fabric comprising the load-bearing elements of structure in a building is able to withstand the effects of fire for an appropriate period without loss of stability. The other three are matters of detailed design which can follow once the adequacy of the structural fabric in terms of fire resistance has been established. Whilst there are examples historically of earth-suspended floors and roofs, our focus here will be on earth load-bearing walls, and how they can meet the required fire resistance standards. The fire resistance of earth walling needs to be assessed in terms of its ability to withstand the effects of fire as follows: (1) resisting collapse, (2) resisting fire penetration and (3) resisting the transfer of excessive heat.

The actual periods of fire resistance required vary according to such factors as building type, use and height. For example, an average two-storey detached house will require 30 minutes of fire resistance for its structural elements, whereas the structural elements of a building used for assembly or recreation and having a height of more than 30 metres will be required to have a minimum period of fire resistance of 120 minutes. A schedule showing the minimum periods of fire resistance for a range of buildings can be found at Table A2, Appendix A of Approved Document B. Generally, Appendix A, of Approved Document B is concerned with the performance of materials and structures, and, amongst other things, includes definitions of 'non-combustible materials' and materials of 'limited combustibility', together with reference to situations where such materials should be used. Included within the definitions of non-combustible materials are those materials which contain not more than 1 per cent by weight of organic material. Hence, most traditional cob-type mixes would not satisfy this definition, as usually they contain more than 1 per cent of straw by weight. However, in German standard DIN 4102 Part 1, earth, even when it contains straw, is considered as non-combustible if its density is not less than 1,700kg/m^3 (Minke, 2000).

In cases where concern is expressed concerning the combustibility of earth mixes containing significant amounts of organic, fibrous material, a sample of the material proposed could be tested, in accordance with BS 476 Part 4:1970 (Non-combustibility Test for Materials) and Part 11:1982

(Method for Testing the Heat Emission from Building Products). Non-combustibility in itself may, for many situations, not be an issue. More often, it will be the period of fire resistance of the wall and the resistance of its surface finishes to the spread of flame which will be critical factors. Again, BS 476 provides guidance on how to test and classify fire resistance and surface spread of flame. Also, guidance is given in BS 6336 'Guide to Development and Presentation of Fire Tests and their Use in Hazard Assessment'. It may be noted, in this context, that tests carried out in Australia on 300mm thick fibreless rammed earth walls gave a fire resistance of four hours, for structural adequacy and integrity (Dobson, 2000). It seems likely, therefore, that a traditional cob wall between 400 and 600mm thick is likely to have a fire resistance of at least two hours, and for most small to medium-sized projects this is likely to be more than adequate.

Experience shows that earth walling generally has a good capacity to resist the effects of fire. Whilst there have been collapses of earth walling during or following a fire, the problem is not usually due to inherent inadequacies in the earth walling itself. More often, the cause is due to lateral restraint being lost or excessive loading being applied from other elements following the burning and collapse of unprotected structural timbers. Further, the action of putting out the fire can have adverse effects. Saturating earth walls through hosing can result in rapid cooling which causes cracks to appear and allows moisture to penetrate. As has been explained previously, too high a moisture content will cause the earth walling to lose its structural integrity, and ultimately will result in collapse.

Approved Document C – site preparation and resistance to moisture

Part C of the Building Regulations is concerned with the preparation of the site (C1), the avoidance of danger to health and safety caused by substances found on or in the ground (C2), the provision of adequate subsoil drainage (C3), and the means of ensuring that the walls, floors and roof of the building adequately resist the passage of moisture to the inside of the building (C4).

Essentially, the focus here needs to be on C4, as C1, C2 and C3 are not concerned directly with the actual fabric of the walls and other structural elements of the building. As far as walls are concerned, Approved Document C advises that the requirements of C4 will be met if the wall prevents undue moisture from the ground reaching the inside of the building, and, if it is an outside wall, it resists the penetration of wind and snow to the inside of the building. Primarily, this is because dampness in a building can be damaging to the fabric and detrimental to the health of the occupants.

Also, the decorative finishes will deteriorate and become visually unattractive. As shown already, excessive dampness in the base of an earth wall ultimately could result in collapse. Approved Document C does not give guidance on preventing damage resulting from the condensation of water vapour on cold surfaces or interstitially, this being a matter for Approved Documents L1 and L2.

It is perhaps worth noting here that while C2 is concerned with dangerous and offensive substances, this is in the context of making the site safe to take the proposed building. However, traditionally materials for earth walling have been obtained close to or from within the site to be developed. Hence, if within a project it is proposed to continue this tradition, and it has been found that the site contains contaminants, it is important to ensure that the material selected for the walling does not contain these. This is a matter dealt with under Regulation 7 in terms of ensuring that the material is suitable and safe for its proposed use.

Historically, earth walling did not have impervious damp-proof courses. Usually in European examples it was built on an underpin course of stone or brickwork which gave it some protection from ground water and splashing. All the constructional materials and finishes of the walls and floors were vapour permeable to an extent, and reliance was placed on a cyclical process of absorbing and releasing moisture to enable the interior to remain adequately dry. Moisture would be drawn into the walls and floor, but usually, through being able to evaporate during dry weather, did not result in an unacceptable build-up of moisture internally. However, with the requirement these days for draught-free interiors and higher internal temperatures, it is more difficult to achieve such a balance. In new-build situations, therefore, compliance with the Building Regulations against rising damp normally would be achieved through the inclusion of a damp-proof course, just as in masonry and brickwork buildings. This should be inserted in the underpin course or plinth rather than in the earth walling itself. As with modern conventional construction, the damp-proof course should be lapped with and sealed to the damp-proof membrane of the solid floor. This approach must relate only to new-build situations. It is never appropriate to insert a chemical or sheet material damp-proof course in an existing earth wall, for reasons which are explained in the second part of this book.

In terms of resistance to precipitation, many modern buildings are based on cavity walls of blockwork or brickwork. Historically, earth walls have relied upon their thickness, often together with an exterior surface coating, to resist penetrating moisture. Also, the common use of generously projecting eaves aids the protection of the wall face. Approved Document C, at Diagram 14, shows an example of a solid external wall rendered externally to achieve a severe exposure rating. However, this example relates specifically to brickwork and types of blockwork, specifying brickwork of at least 328mm thick and dense aggregate blockwork of at least 250mm thick. While within the Approved Document there is no specific reference to unbaked earth walls, it is reasonable to assume that they can be made to meet an acceptable standard, since usually external walls of earth would be considerably thicker than the blockwork or brickwork examples included. Rendering for the examples given must be in two coats with a minimum total thickness of 20mm. A rendering mix is given, but this is comprised of 1:1:6 cement/lime/sand, which would not be appropriate for earth walling. Hence, reliance may have to be placed here on historic precedent, showing that similar constructions in similar environmental conditions, but with lime- or earth-based plasters, have proved to be adequate in the past. Appropriate finishes for earth walls are fully described in Chapter 4.

Approved Document E – resistance to the passage of sound

Part E of the Building Regulations deals with sound insulation, and a revision to this took effect in July 2003. The schedule to the earlier regulations stated requirements for airborne sound (E1 – walls; E2 – floors and stairs) and for impact sound (E3 – floors and stairs) relating to the separating structure between a dwelling and another dwelling or part of the same building. The 2003 revision removes specific reference to airborne and impact sound within the regulations themselves. However, Tables 1a, 1b and 2 of the accompanying Approved Document E lay down required standards for airborne and impact sound insulation for floors and airborne sound insulation for walls in relation to specific situations. The key differences in the revised regulations are the additional requirements for reasonably resisting sound penetration between certain rooms within a dwelling (whether internal walls or floors), reducing reverberation in the common internal parts of certain residential buildings to acceptable levels, and achieving acceptable acoustic conditions and sound insulation in schools. There is recognition at paragraph 0.6 of the Approved Document that in the case of some historic buildings undergoing a material change of use, it may not be practical to improve the sound insulation to the standards set out in the tables. In such cases, the sound insulation should be improved to a level that can be achieved practically without prejudicing the historic character of the building.

In terms of new-build, for sound insulation earth walls should not present any more of a problem than masonry materials. Largely, it is the mass of a wall which determines its resistance to airborne sound. Cob usually has a density in the region of 1,800kg/m^3 and for a wall 450mm thick this will give a mass of 810kg/m^2. This compares favourably with the alternatives of brickwork, blockwork and *in situ* concrete.

Approved Document L1 – conservation of fuel and power in dwellings

Approved Document L2 – conservation of fuel and power in buildings other than dwellings

Before going on to discuss in detail the problems that need to be overcome in order to comply with Parts L1 and L2 of the Building Regulations, it might be useful to describe, in outline, the thermal characteristics of mass earth walls. According to Houben and Guillaud (1994) in mass earth walls, 'the relationship between the coefficient of thermal conductivity, or "k" value in W/mK, and the bulk density of the material (kg/m^3) is very close to that of other mineral materials', fired brick and concrete blocks, for example. The authors then go on to say that, 'Even though soil has a lower thermal capacity than other heavier materials, its ability to store heat, which logically speaking should be lower, is nevertheless excellent [because] soil benefits from a latent inertia related to its absorption capacity.' This property, which is sometimes referred to as thermal inertia is, however, not something that the Building Regulations take into account; the principal factor applied in assessing the thermal performance of natural (non-manufactured) materials being thermal conductivity based upon measured dry density. A rough guide to how conductivity relates to density in earth-based materials is shown in Fig. 5.1.

The wide variation in the densities of various naturally occurring soil types, mentioned in Chapter 2, has been shown to range from around 1,500kg/m^3 in chalks and clayey silts up to 1,900 to 2,000kg/m^3 in gravelly sands. In theory at least, it should be possible to reduce a soil's thermal conductivity by reducing its density; and this is true, but only up to a point. As already noted, the density of, for example, a gravelly sand soil can be reduced by (a) screening out gravel and stones over, say, 10mm diameter and (b) adding to it the maximum possible amount of straw during the mixing process. (This would only be possible if one of the 'wet' methods of construction were being employed.) Density could be more significantly reduced by the addition of lightweight aggregates. However, in doing this the compressive strength of the material might be reduced to the point where it was no longer suitable for mass earth, load-bearing construction.

One organic, renewable material that might have some potential in this respect is sawmill waste, or softwood shavings, formerly a waste

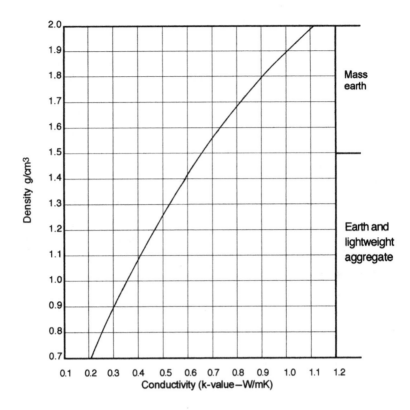

5.1
Relationship between density and thermal conductivity in earth-based building materials
Adapted from Minke (2000)

product but now used in the production of particle board for the building industry. Some preliminary testing carried out using a mix of one part clayey sand soil to two parts of softwood shavings, by volume, would suggest that a density of around 1,100kg/m³ could be achieved, thus giving a thermal conductivity (k-value) of 0.41W/m²K. If the material were tamped into formwork (it would not be suitable for the wet, piled form of construction) a compressive strength of about 650kN/m² would be achieved, quite sufficient to support a 500 to 600mm thick single-storey wall, or a two-storey wall if excessive point loading could be avoided. A 600mm thick wall, with lightweight renders and plasters applied externally and internally, would give a theoretical U-value of around 0.56W/m²K. However, the material could clearly not be regarded as non-combustible and its fire resistance would no doubt be much lower than that of mass earth. Clearly, there is a need for further research in this field, and some preliminary work has been carried out at the University of Plymouth, where a method has been developed for measuring and recording, *in situ*, the thermal conductivity and diffusivity of mass earth walls and associated materials (Goodhew *et al.*, 2000).

The principal challenge in conforming to the Building Regulations is in satisfying the requirements for the conservation of fuel and power, as covered by Parts L1 and L2. At the time of writing, the 2002 editions of the Approved Documents L1 and L2 are applicable. However, revised standards are imminent, and these will require improved standards of energy conservation. Approved Document L1 (dwellings) shows three methods of showing compliance with the 2002 Regulations, namely (a) Elemental, (b) Target U-value and (c) Carbon Index. Approved Document L2 (Buildings other than Dwellings) also shows three methods, these being (a) Elemental, (b) Whole-building and (c) Carbon Emissions Calculation.

For both dwellings and other building types, the Elemental Method is the simplest in terms of calculation, and defines maximum allowable U-values for the various building elements. The Target U-value Method for dwellings is more flexible than the Elemental Method in that it allows a certain level of trade-off between the various building elements. It also takes into account the efficiency of the heating system and enables solar gain and building orientation to be taken into account.

Similarly, in non-domestic buildings the Whole-building Method allows more flexibility than the Elemental Method. It is concerned with the performance of the whole building, thereby enabling a level of trade-off between the elements, and for offices requires that the Whole-office Carbon Performance Rating meets specified standards set down within the Approved Document. The Whole-building method was used to calculate the heat loss of the external envelope of the Norden Reception Building, Corfe Castle, Dorset, constructed during 1998 (see Fig. 5.2 and Colour Plate 6). By the use of a very highly insulated plinth and roof, an acceptably low level of heat loss was achieved. For schools and hospitals, the Approved Document refers to specific documents which lay down the required standards.

The Carbon Index Method and the Carbon Emissions Calculation Method allow even more flexibility. The Carbon Index Method requires compliance with the Government's 'Standard Assessment Procedure (SAP) for Energy Rating of Dwellings' (*2001* edition at time of writing). The Carbon Emissions Calculation Method requires that the calculated annual carbon emissions of the proposed building should be no greater than those from a notional building of the same size and shape designed to comply with the Elemental Method. Unlike the Elemental Method, it enables advantage to be taken of any energy conservation measures, and takes account of useful solar and internal heat gains.

For dwellings, the Elemental Method is seen as suitable for alterations and extension work, and for new-build when it is desired to minimise calculations. However, it may be difficult to apply this method to buildings having earth walls due to the required maximum U-value for walls of $0.35W/m^2K$. In Britain, there are no formally accepted standards of thermal transmittance for earth wall types, since the actual constituent materials vary

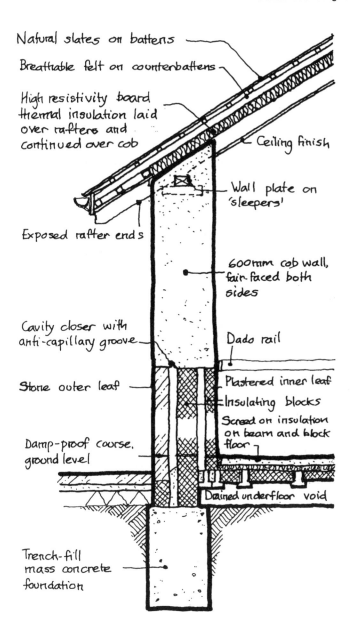

Natural slates on battens

Breathable felt on counterbattens

High resistivity board thermal insulation laid over rafters and continued over cob

Ceiling finish

Wall plate on 'sleepers'

Exposed rafter ends

600mm cob wall, fair-faced both sides

Cavity closer with anti-capillary groove

Dado rail

Stone outer leaf

Plastered inner leaf

Insulating blocks

Screed on insulation on beam and block floor

Damp-proof course, ground level

Drained underfloor void

Trench-fill mass concrete foundation

5.2
Norden Park-and-Ride reception building; typical section through external wall

widely. It should be noted that, in using the Target U-value and Carbon Index Methods to comply with Approved Document L1, the highest (worst) allowable U-value for external walls is 0.7. In order to achieve this a mass earth wall, of density 1,700kg/m³, would need to be 1.15 metres thick. However, a 600mm thick wall constructed of the same material, but with an added layer of 50mm insulation, and normal lime- or earth-based plasters and renders applied, would achieve a theoretical U-value of 0.45 – well within acceptable limits.

Rammed earth is of higher density than cob, and is usually associated with thinner walls. Hence, its U-value is likely to be even more removed from the required standard. In most building projects today it would be impractical to make a wall totally of earth sufficiently thick to meet the required standard, so other means of doing so need to be considered. There are a number of options for this, one of which involves using the earth walling in combination with other materials. Historic precedents exist: for example, cob or clay-lump walled buildings with a facing of brickwork. Sometimes such brickwork facings are a later modification, but it is believed that during the late eighteenth and early nineteenth centuries on many buildings the brickwork and cob were built concurrently.

Without resorting to excessive thickness, today it would be necessary to use materials of much greater thermal insulation value than brick in association with the earth walling. In doing so, it is necessary to balance a number of possibly conflicting factors. For example, the use of some plastic-based thermal linings may be considered to be inappropriate through their high embodied energy, and therefore questionable upon sustainability grounds. Alternatively, there may be concern at using factory-produced facing materials which detract from the visual plasticity of form associated with some earth-walling types. If the desire is to achieve this 'organic' appearance externally, then it is possible to apply the thermal insulation lining internally so that the thermal requirements of the Building Regulations can be met. However, such a method limits the potential for gaining internal benefit from the thermal capacity of the earth walling, since heat gain in the structure cannot pass readily to the inside. Internal thermal linings are generally considered more effective when associated with lightweight construction, such as timber framing, and rapid response heating systems.

Conversely, mass earth walls lend themselves more readily to buildings where a more or less constant internal temperature is desired, and where the thermal storage and radiation capacity of the walls can help prevent rapid temperature change in the internal space. In order for this process to work efficiently any additional insulation required to minimise heat loss through the wall to the outside air needs to be applied at or near to the external wall face. In achieving this, however, one needs to reconcile two potentially conflicting requirements: the need to allow free movement of water vapour from the wall while, at the same time, ensuring that penetrating dampness does not adversely affect its thermal transmittance or U-value. Any finish applied to an external wall will also need to be reasonably resistant to erosion as a result of weathering, and to abrasion and impact damage.

Two possible solutions might be (1) to apply studwork and timber cladding to contain and protect the insulation layer or (2) to fix the insulation layer to the earth wall and then apply a 20mm thick lime/sand or lime/earth render on to some form of mesh or lathing in order to stiffen and reinforce it. The first solution may be expensive in both material and labour costs and, in areas where there is no tradition of weatherboarding or timber framing, may be objected to on aesthetic grounds. It would also, of course, completely obscure the earthen nature of the construction. The second solution would require the use of some form of non-corrodible metallic reinforcement and secure fixings into the earth walls but, if earth were to be used in the finishing render coat, would more closely resemble the mass earth wall to which it had been applied, although it would lack the gently undulating wall surfaces and rounded corners characteristic of traditional cob walls.

Nevertheless, there are recent examples of external walling comprising an earth mix with an internally applied thermal insulation layer. David Glew has employed this technique on extensions to mud-and-stud houses in Lincolnshire, England. In these cases, the extensions follow the local tradition of incorporating a timber frame and vertical laths within the depth of the wall. The internal insulation is a proprietary composite board which includes 41mm of rigid phenolic foam, and the U-value achieved for the total wall thickness is 0.347 (Glew, 2000).

One example of a separate insulation layer being used externally is the recently constructed shop and information centre at the Centre for Alternative Technology, mid-Wales, where rammed earth walling has been constructed behind a highly insulated external timber stud wall. The reason for this was not only the need to achieve the required thermal insulation level, but also a concern about exposing an unprotected earth wall to the wet Welsh climate. Other examples include the application of an insulation layer to the external face of the earth wall, which in turn is covered by a facing material such as timber boarding or clay tiling.

The incorporation of vertical insulation boards into the core of a cob wall during construction is not considered to be practical. However, some earth-walling types lend themselves more readily than others to composite construction incorporating a thermal layer within their overall thickness rather than internally or externally. In one recent rammed earth project, for example, 50mm thick rigid foam insulation was incorporated within a 450mm thick wall section during construction, with wall ties placed through the insulation to connect external and internal skins of the wall. In another rammed earth project a U-value of 0.7 was achieved by applying a 60mm thick lime-based external render coat, containing pumice stone aggregate (University of Bath, 2002–4). Mud bricks and blocks can be used for cavity external walls, in which the thermal insulation is located within the cavity zone. The process would be very similar to that currently in common use for masonry cavity walls. For example, a wall constructed with an outer skin of 150mm earth/straw blocks (density 1,700kg/m^3), a cavity filled with 50mm insulation, and an internal skin of 225mm 1:2 earth/softwood shavings blocks (density 1,100kg/m^3) with a 20mm lightweight plaster internally and 10–15mm external lime/sand render, would achieve a theoretical U-value of 0.40. Some clay lump walls incorporating an insulated cavity have been built in Norfolk, and more are planned (D. Bouwens, pers. com.).

In order to reduce aggregates, the thermal conductivity of a mass earth wall, the incorporation of lightweight additives may be considered. In making such a decision, account needs to be taken of the other required performance standards of the walling material, such as load-bearing capacity, sound insulation and fire resistance. Apart from softwood shavings, which have already been mentioned, other organic materials, such as straw, reed, wood-wool and cork may reduce all of these other qualities. Lightweight mineral additives, such as perlite and expanded clay, will significantly reduce the load-bearing capacity and possibly the level of sound insulation. What is more, they are products of combustion, and therefore do not have the sustainability credentials of earth-walling materials which are dug, mixed and used locally in their raw state. Nevertheless, in small-scale projects where the load-bearing capacity and sound insulation requirements are not unduly restrictive such additives may be a practical solution.

Enhancing the thermal insulation qualities of earth-walling mixes is a subject currently receiving extensive research (Goodhew and Griffiths, 2005), and it is hoped that the ready access of nationally accepted data showing how statutory standards can be achieved is imminent. However, even without enhancing the thermal resistance of the earth walling itself, compliance with the Building Regulations Parts L1 and L2 is possible. It can be achieved through using the Carbon Index and Carbon Emissions Calculation methods, following a holistic approach which takes into account all the elements of the external envelope, the heating system and the orientation of the building. In addition, there is considerable scope for using earth walling in situations where achieving mandatory U-values is not a critical factor. For example, earth walling could be used internally in association with external glazed areas such as conservatories, enabling heat to be stored within the earth walling and maximising the benefits of its high thermal capacity. This 'passive solar' type of design solution might present the opportunity to mix ancient and modern technologies in a visually exciting way while at the same time adhering to ecological principles. Other opportunities exist for using earth in ancillary structures. For instance, material dug to accommodate foundations and drainage sometimes would be usable in earth wall construction, perhaps for boundary and sound buffer walling.

The forthcoming amendments to the Building Regulations dealing with conservation of fuel and power need not necessarily impose greater restrictions on the use of earth external walling. Indeed, it is likely that they will give more latitude by allowing greater scope for achieving better energy conservation by means other than prescribed U-values. In July 2004, the Office of the Deputy Prime Minister published proposals for the next major revision of energy performance in buildings, setting a target of 25 per cent improvement in energy savings. The indications are that for domestic work, the Target U-value Method and the Elemental Method will be scrapped in favour of Government-approved Standard Assessment Procedure (SAP) ratings. The Standard Assessment Procedure defines the means of producing an energy cost rating (the SAP rating) and a carbon index for a dwelling, based on calculated energy for space and water heating. The new ratings will have to satisfy the demands of the EU (European Union) Energy Performance Directive. For non-dwellings a 'national calculation methodology' is to be introduced, again in accordance with the requirements of the European directive. Increasingly, building control compliance will rest with the use of approved calculation software, versions of which have been developed already by the Building Research Establishment. This should give applicants greater scope to claim compliance through the submission of calculations prepared by suitably qualified professionals. Increasingly this is likely to result in model design packages prepared by trade associations and product manufactures, and perhaps there is an opportunity here for a body representing earth-walled building designers and builders to become proactive in this area.

In summary, statutory standards need not be an impediment to the use of earth walling, and this is borne out by the evidence of its resurgence in some European countries, America, Australia, New Zealand and other countries where its traditional use had virtually died out. Designers and builders in these countries have shown that the perceived problems of complying with statutory controls can be overcome, and have played their parts in bringing about an increasing use of earth walling. It is to be hoped that a growing recognition of the need for sustainable construction and the broad range of factors to take into account in achieving it will prompt changes in the regulations of those countries where statutory controls are seen currently as inhibiting the use of earth walling.

Technical standards and codes of practice for earth-wall construction

In some countries, where earth has been used extensively in recent years for the construction of new buildings, technical standards and codes of practice have been adopted and now form the basis of regulatory or advisory systems for controlling and overseeing earth-wall construction. Countries whose national or, in some cases, regional governments have now adopted such standards include Germany (DVL, 1999), New Zealand (NZS 4297–4299, 1998), Australia (Standards Australia, 2002), Switzerland and the USA. Both the German and New Zealand systems appear to be comprehensive and detailed, which is helpful when one is dealing with forms of construction that represent what might be termed the 'engineered' approach, such as mechanised rammed earth and compressed soil blocks, which are fairly easy to control and regulate. In other, more traditional forms of earth construction, however, such as cob and mudwall building, where the approach is rather more intuitive and dependent upon the skill of the individual craftsman, regulatory controls may be very difficult to apply. Although the types of soils to be used in these forms of construction could be specified, their suitability assessed by means of approved test procedures, and recommended codes of practice agreed to, strict on-site quality controls and compliance with agreed standards would be difficult to enforce. Great reliance therefore would need to be placed on the work being supervised by experienced and qualified practitioners.

At present there are no British standards relating to the use of earth for building construction, and this seems likely to remain the case, at least in the short to medium term. However, it seems clear that interest in sustainable building is growing and that unbaked earth as a 'green' building material has much to commend it. Given that new earth construction in Britain is still at a fairly early stage of development, and that there is a dearth of practical experience in using the material for building, then clearly the present need is for guidance rather than for control or regulation. In setting standards and codes of practice for earth building there are two key issues that need to be addressed. The first relates to the raw material and how its suitability for building construction may best be assessed, a topic that has been discussed at some length in Chapter 2, and the second concerns the need to ensure that the construction process is carried out to a satisfactory standard by suitably trained and experienced personnel. For the 'engineered' earth-building methods, technical standards and codes of practice that already exist in the countries mentioned above could probably be adapted for use in Britain. At the time of writing, a research project entitled 'Developing Rammed Earth Walling for UK Housing Construction' being coordinated by the University of Bath, is nearing completion. One of its principal aims is 'to produce design and construction guidance notes for rammed earth wall construction' intended to persuade the building industry that this non-indigenous technique represents a viable and sustainable form of building construction (Walker *et al.*, 2005).

Does all this, together with the increasingly onerous thermal performance requirements enshrined in the Building Regulations, signal the imminent demise of centuries-old traditional building methods such as cob, mudwall and clay lump? The answer must be an emphatic 'no'; both new and traditional methods of earth construction must continue to co-exist, for the following reasons: While accepting that the 'engineered', semi-mechanised forms of earth construction are the only ones likely to be acceptable for volume house construction and for public building projects, for the self-builder or for community building projects based on the use of volunteer labour, the wet, piled or clay lump methods have much to commend them. Also, for someone seeking to build an 'organic', hand-crafted, uniquely individual house, cob or mudwall construction is probably the only answer. Finally, and perhaps most importantly, traditional craft skills such as cob and mudwall building need to be kept alive in order to repair and maintain the rich and diverse heritage of surviving earth buildings in Britain.

Part 2

The conservation, repair and maintenance of earth buildings

Introduction

Historical background – public perception of earth buildings

In many parts of the world earth construction represents part of a continuous centuries-old tradition. In most of Western Europe, however, as noted in Chapter 1, this is clearly not the case. Up until the mid-nineteenth century earth buildings were commonplace in many parts of Britain. In some areas earth buildings were regarded as being of low status, suitable only for the poor, who lacked the means to build more durable houses of brick or stone masonry. However, in other areas, most notably south and south-west England, earth was used, from the late medieval period onwards, to construct buildings for and by people from a wide social spectrum, from the very poorest landless labourers up to the most prosperous yeoman farmers and minor gentry. It is probably for this reason that many more earth buildings have survived in this area than in most other regions of Britain. Nevertheless, even now in south-west England the public perception of earth buildings is generally unfavourable, and most people would be very surprised to learn that thousands of such buildings still exist in Britain.

Hopefully, the preceding chapters, especially Chapters 2, 3 and 4, will have helped to dispel the idea that well-constructed earth buildings are in any way inferior to those of masonry or lightweight-framed construction and are, in fact, in many ways superior in terms of their low-energy characteristics and their provision of a healthy living environment.

The survival rate of earth buildings is dependent, among other things, on their structural soundness, which itself depends upon the quality of materials and workmanship employed in their original construction. As an example, one might compare two English counties – Lincolnshire and Devon. The mud-and-stud buildings of Lincolnshire are, compared to the cob buildings of south-west England, rather insubstantial and consequently more susceptible to decay and damage from the relatively harsh climate of this part of England. It is therefore not surprising that, of the thousands of mud-and-stud buildings known to have existed in the county in the nineteenth century, only a few now remain, and for this reason strenuous efforts are being made to conserve them (Hurd and Gourley, 2000).

During the nineteenth and early twentieth centuries thousands of earth buildings were either pulled down by 'improving' landowners, often in collusion with the Parish Overseers of the Poor, or, at a later stage, effectively legislated out of existence, usually, but not always, for the best of reasons – to improve the living conditions of the 'labouring poor'. The Reports to the Board of Agriculture, produced for every county in England and Wales, and published between 1793 and 1815, often make for grim reading. For example, Arthur Young, in his 1808 report on Sussex, wrote that 'the dwellings of the poor are, in most counties, but mud cabins [sic], with holes that expose the inhabitants to the rigour of the climate', while Charles Vancouver, in his report on Devon published in 1807, stated that in the western part of the county, 'three mud walls and a hedge bank form the habitation of much of the peasantry'. Vancouver, in common with other visitors to Devon, was less than enthusiastic about the appearance of the poorer cob buildings, and deplored the practice of leaving their walls unrendered because they were indistinguishable from the fields surrounding them, having the appearance of what he termed 'rude, primitive huts'. (Of course, the idea of integrating buildings into the rural landscape is one that would be regarded as highly desirable today.) Forty years later, a correspondent writing to the government about rural housing in Dorset described 'the general condition of the cottages of the peasantry' as 'miserable, deplorable, detestable, a disgrace to a Christian community' (Machin, 1997). That conditions had improved very little by the third decade of the twentieth century, when British agriculture was suffering a severe depression, can be seen from reports submitted to the Government by County Medical Officers of Health in 1926, one of which, relating to Devon, had the following to say:

> The village can best be described as a slum. I find it impossible to describe adequately in words the primitive squalor that obtains. They [the working-class houses examined] ought to be closed and demolished without delay, as they are liable to collapse at any moment.

Partly as a result of this and many other alarming reports, the Government quickly passed legislation, in the form of The Housing (Rural Workers) Act 1926, which enabled local authorities to provide grant aid towards the rehabilitation of rural cottages. By the outbreak of war in 1939, 2,082 rural dwellings, probably half of which were of cob construction, had been reconditioned in Devon, while in the much smaller, less populated county of Cumberland (now part of Cumbria) 1,103 houses had been similarly improved (Shears, 1968). Although, clearly, this scheme was introduced primarily for social and economic reasons, there was, nevertheless, tacit acceptance of the need to respect the character and appearance of the rural environment, by making grants conditional on external building works conforming to local vernacular building styles. Thus were the first tentative steps taken on the road to vernacular building conservation, and the value of earth buildings, in social and economic terms as part of the national housing stock, formally recognised. As for the many thousands of earth buildings that vanished more or less without trace during this period; they presumably returned to the ground, from whence they came.

Why preserve earth buildings: the need for conservation

The most compelling reason for preserving any house is that it represents, in financial terms, a valuable and appreciating asset. For example, a farmhouse built in the seventeenth century for, say £25 to £30, would, depending on its location, in today's market be worth between £250,000 and £400,000, irrespective of whether its walls were of earth or masonry construction. Even fairly humble terraced cob and thatched cottages located in rural areas command premium prices. However, agricultural, industrial and other nondomestic buildings are quite another matter. Where, occasionally, an ancient, evolved farmstead survives with all its traditional buildings intact and with some or all of its earth or chalk walls exposed, sited for example in a wooded valley, this can present an extraordinarily picturesque and evocative scene; the buildings giving the impression of having grown organically from the land on which they stand and being perfectly integrated into the rural landscape. However, sentiment and nostalgia have little part to play in modern agriculture and, in fact, many of these ancient farmsteads have been almost entirely destroyed, with only the farmhouse left standing, surrounded by steelframed, corrugated asbestos or metal profile sheet-clad, raw concrete block buildings. The reasons for this apparent wholesale vandalism and despoliation of the rural environment are quite clear. Most traditional farm buildings, even if they could be adapted to modern farming practices – which mostly they can't – are very expensive to maintain properly. So, it's all a question of practicality and economics, as most farmers would probably agree. Clearly, of all the earth buildings that still survive, it is the redundant farm building that is most at risk. In practice, even statutory protection is no guarantee of survival and large numbers of these structures are lost every year, their demise often unnoticed and therefore unrecorded. Some redundant barns are 'saved', with official sanction, by being converted for residential use. Unfortunately, this often results in extensive demolition followed by rebuilding in concrete blockwork. Probably the only circumstances under which a use could be found for these traditional farm buildings would be where the small-scale, intensive, organically based farming practices of former times were being employed. (Some older farmers have been heard to remark that animals were healthier and probably happier when they were housed in well-maintained cob buildings, especially those with thatched roofs.)

Statutory protection

In England, Scotland and Wales, buildings officially classified, or 'listed', as being of historic or architectural interest have, at least in theory, the benefit of statutory protection from demolition or inappropriate alteration, as do their ancillary buildings or other 'curtilage' structures such as boundary walls. A limited degree of protection is also provided to unlisted buildings located within designated 'conservation areas'. Although quite large numbers of earth buildings are listed, this is not normally because they are constructed of earth but because they meet the required criteria, which are based primarily on either a building's intrinsic architectural merit or the contribution it makes, in visual terms, to the character and appearance of a town or village street scene. It is for this reason that the vast majority (probably more than 80 per cent) of surviving earth buildings lack any form of statutory protection.

The case for conserving earth buildings can be made, firstly, on socio-economic grounds; taking into account the value of such buildings to their owners and to the nation as part of the total housing stock. Secondly, earth buildings should be valued for the way in which, when well maintained, they are able to provide a healthy and comfortable living environment for both people and animals. Thirdly, and perhaps most importantly, consideration should be given to the individual 'hand-made' character of earth buildings, even those considered to be of little or no architectural merit, and the important role they can play in defining and enhancing a unique local identity or 'sense of place', particularly in the countryside, and in rural towns and villages.

Having, hopefully, justified and explained the need for conserving traditional earth buildings, the final two chapters of the book will describe the ways in which this can best be achieved: by identifying the principal causes and effects of various types of building failure, by explaining how to repair earth walls in the least disruptive, most sensitive way, and by suggesting ways in which their durability and longevity can be ensured by means of appropriate maintenance. Both chapters are concerned primarily with the repair, conservation and maintenance of buildings constructed using the traditional methods most common to Britain – cob, mudwall and clay lump. The reason for this is that the rammed earth technique has rarely been employed here, other than in the rammed chalk buildings of Wessex, which are relatively late in date (mid-nineteenth century) and few in number. Clearly, therefore, very little experience has been gained in dealing with structural failures in buildings of this type. However, despite this, it is considered that some of the general principles and repair techniques described might also be appropriate for use in rammed earth and similar forms of mass walling.

Chapter 6

The principal causes of failure in earth walls and how to recognise them

Inherent defects associated with original construction methods and materials

Walls, plinths and foundations

It should be evident, at this stage in the book, that the durability and longevity of an earth wall will be dependent as much upon the raw material originally employed as it is on the degree of care with which the building was constructed. It would perhaps be reasonable to assume that few buildings that were built in haste, with little care and using unsuitable materials, still survive. However, experience has shown that such buildings do exist, though they are often difficult to identify until they develop major problems. This is particularly true of cob and mudwall buildings because such great numbers of them were built over a period of at least seven centuries, whereas virtually all clay-lump buildings date from the nineteenth century or later, and very few rammed earth buildings, apart from the chalk *pisé* buildings of Wessex, also of nineteenth-century date, have been identified in Britain.

If the soils used to construct a mass earth-walled building were deficient in some way (if, for example, they contained insufficient clay, combined with either excessive amounts of stone and gravel or with large quantities of silt and other fine material) then they will lack the compressive strength, and resistance to the effects of excess moisture, of a well-graded, clay-rich soil. However, the presence of a large clay fraction – in excess of 20 to 25 per cent – can bring its own problems; those of swelling and shrinkage, which, unless they are controlled, may lead to the development of cracks and fissures as well as excessive wall settlement (vertical shrinkage) during the construction and subsequent drying-out process. Where an unmodified clay-rich soil, containing insufficient organic fibrous material (usually straw) has been employed and the material has not been well compacted, cracks and fissures, either actual or incipient, will probably have developed at key points in the structure, in particular, at internal corners, and above and around door and window openings. Other potential weak points, hollows and cavities may be the result not only of inadequate compaction but also of poor distribution of the straw binder. Concentrations of straw within the wall mass and between lifts, where layers of straw were often used protect the cob from rainfall while under construction, can also form points of structural weakness as well as potential nesting sites for rodents and, in unrendered walls, for birds and masonry bees.

In traditional, manually rammed, *pisé* walls, which are inherently stronger than those constructed using the wet, piled method, deterioration and damage are usually confined to the external faces of unprotected walls. Because cohesion in rammed earth walls is achieved primarily by means of surface friction between, and physical interlocking of, the individual soil particles, and because heavy ramming reduces the soil's porosity, such walls are much more resistant to the effects of moisture penetration. However, they are, as already noted, susceptible to surface erosion and frost damage unless protected either by substantial overhanging eaves or the application of a suitably porous render coat.

Foundations and plinths will vary according to the type and date of construction, the status of the building and what raw materials were locally available. For example, in earlier times, especially in remote rural areas, good quality building sand was often difficult to obtain and building lime was probably beyond the financial means of many people. Masonry plinths for cob buildings, therefore, tended to be fairly minimal or even non-existent, and in those that were built the facing stones were often bedded in an earth or earth/lime mortar, and the core of the wall filled with earth and stone rubble. Most usually no provision was made for tying together the external wall faces by the inclusion of cross-bonding stones. Building methods only started to improve significantly during the nineteenth century, when improved transport links – turnpike roads, canals and railways – facilitated the movement of building materials such as fired bricks, quarried sand and stone, and building limes. Poorly constructed masonry plinths and lack of proper foundations are both potential sources of problems, particularly in cob walls, for two main reasons. Firstly, earth and weak lime mortars may decay and then be leached out of joints in the wall face, allowing the stones to move and the wall to bulge out. When this happens, loose rubble in the core of the wall will drop down to fill the resulting cavities, thus exerting further outward pressure on the facing stones and causing the plinth to become unstable. Although the cob above will often remain intact, if a substantial enough section of plinth moves, then the cob wall, being no longer fully supported, might start to tip and move outwards. Secondly, where the foundation, made up of separate stones rather than a continuous strip of concrete, has been laid on a clay-rich, expansive subsoil, differential settlement may occur as a result of seasonal wetting and drying out of the area immediately below the foundation. This can result in a similar situation to that described above except that the outer wall face will sink rather than bulge, again leaving the cob above partially unsupported and leading to the development of a vertical crack, or split, in the core of the wall.

Other weak points in cob buildings are associated with chimney stacks and flues. For example, in earlier buildings cob external (projecting) stacks, as well as being susceptible to rising and penetrating dampness, can also be affected by the differential ground settlement referred to above. In later, nineteenth-century buildings, chimney flues were often formed within the thickness of the wall, thus creating structural discontinuities and forming points of weakness in otherwise monolithic walls.

Roofs, carpentry and other building elements

When serious structural failures occur in cob, mudwall and clay lump buildings it is not always possible to attribute the failure to one single cause. Most usually, the failure has occurred as a result of several factors acting in combination. Although an earth building is made up of discrete structural elements – foundations, external walls, roof, chimney stacks, internal walls and floors – in earth-walled buildings all these elements are structurally interdependent so that, in seeking to identify the causes of decay and structural failure, the whole building needs to be thoroughly examined. Serious structural failures in cob and mudwall buildings are much more likely to occur in later, eighteenth- and nineteenth-century buildings than in those dating from the medieval and early modern periods, up to around 1650; though it is probably true to say that most buildings that do survive from the earlier period tend to be houses of higher status and were, therefore, well constructed. In the earliest domestic buildings, which tended to be hall-houses with no upper floor, other than perhaps a solar at one end, the earth walls were usually quite thick, 700mm or more, and were often no more than three metres high, giving a slenderness ratio (width to height of wall) of around 1:5. Given also that the roof structures were of very substantial oak construction, with the jointed cruck feet of the principal rafters bedded well into the cob walls (see Fig. 6.2) then it may be seen that these houses were built to last, which indeed many of them have, though usually in much altered form. From the mid-seventeenth century onwards, building styles changed, mainly in response to the changing needs and aspirations of an expanding and increasingly prosperous population. Two-storey houses with built-in chimneys and larger windows became the norm, and improvements in agricultural practice and productivity required a range of larger, more specialised buildings. During this period – up to the mid-nineteenth century, apart from the introduction of shuttered cob for some buildings and the occasional use of lime stabilisation, cob and mudwall construction techniques remained much the same as they had been during the medieval period. However, as noted above, the buildings themselves changed quite significantly. Walls became higher but, at the same time, reduced in thickness (typical slenderness ratios of 1:8 to 1:10), while roof structures became more basic, less elaborate because they were no longer on view and, in many cases, reduced to the bare minimum. What is known popularly as the 'A-frame' roof truss is a potential source of problems, particularly when it forms part of a fully hipped roof. The A-frame truss was probably a development (if this is the right word) of earlier cruck and jointed cruck forms (see Fig. 1.1C) but is greatly simplified and much less substantial. It is made up of two principal rafters, the feet of which are normally supported on timber bearer plates bedded and pegged into the cob wall head, joined by a collar tie located halfway up the rafters, sometimes higher. There are no cross-ties at eaves level and no wall plates to support the common rafters, which are carried usually on two sets of purlins and have their feet simply bedded into the cob wall head.

A-frame trusses, because their feet are not restrained by eaves-level cross-ties, and because they lack any form of diagonal cross-bracing, are very susceptible to flexural movement and subsequent failure if overloaded or otherwise weakened. Failure will usually occur at or around the mid-point of the rafter, and this will result in a horizontal force acting on the wall head, thus pushing it out.

Roof types vary according to age and location of building. Regional and local variations include half-hipped roofs or roofs with one end hipped and the other gabled with a built-in chimney stack. Also found are fully hipped roofs with a central (axial) stack, which also acts as a structural support for the roof timbers. Fully hipped roofs were popular in some areas, for two reasons. Firstly, because building up to the apex with cob is a rather difficult, laborious process, for which reason gable walls were sometimes built up with timber framing and finished with lath and plaster. Secondly, hipped roofs are easier to thatch and are much more effective in protecting earth walls from rain damage. The main problem with hipped roof structures is that they are prone to longitudinal movement (known as 'racking') as a result either of failure of the hip rafters or movement in one or more of the main roof elements, creating a 'domino' effect, which will tend to push out the gable end wall (see Figs 6.1A and B).

6.1
Above, corner crack in cob barn, also showing entry holes to 'rat runs' and erosion from water run-off. Below, detail of barn showing outward-leaning gable wall and displaced stones in plinth

During the late eighteenth and nineteenth centuries fashion dictated that earth-walled buildings should appear identical to those of masonry construction. Builders were able to respond to these demands simply by basing their designs and specifications on material contained in the various architectural pattern books and builders' directories that became widely available at this time. Only minor adjustments needed to be made to allow for thicker walls and for fixing timber components into the cob or rammed earth from which they were constructed. Roofs were more or less the same as those of contemporary masonry buildings – fully triangulated and braced, and supported on wall plates. Some houses, picturesque villas and 'cottages ornés', had fully hipped roofs, either slated or thatched, with the deep eaves complete with decorative embellishments fashionable at the time, which also provided protection for the earth walls. Many town houses, on the other hand, were built to the standard late Georgian pattern, with shallow pitched or mansard roofs, their main facades topped with parapets, and with minimal overhang at eaves and verges (see Figs 1.13 and 6.2). In such houses rainwater penetration through defective parapet gutters, and around rainwater downpipes and chimney flashings, pose particular problems, which, if not dealt with, can result in serious weakening of the earth walls and subsequent failure. Other buildings potentially at risk are large structures such as threshing barns, warehouses and other industrial buildings, and, in particular, non-conformist chapels, many of which date from this period. Usually, buildings of this type have relatively slender walls, around six metres or more in height, but lack the internal cross-walls and intermediate floors that would provide lateral restraint to the external walls of a domestic building of similar size. Add to this the loads imposed by wide-span roofs, often with no intermediate support, and it will be realised that such buildings are very susceptible to the effects of structural movement and moisture penetration. In recent years several serious failures, including partial collapse, have occurred in large buildings of this type.

Problems arising from later alterations and interventions

Alterations to roofs

The most common alteration to the roof of an earth building is replacement of the original covering of slate or thatch with some other material. Thatch, usually wheat or water reed, was formerly ubiquitous in the rural areas of Britain and, prior to the mid-seventeenth century, also common in towns and cities. Contrary to popular belief, thatch is not a lightweight roofing material. At 40kg/m^2 for a 300mm thickness, it is about the same weight as medium thickness natural slates or interlocking plain clay tiles. If a thatched roof has

6.2
Evolution of the cob wall. A late medieval building with jointed cruck roof compared with an early nineteenth-century two-and-a-half-storey town house with a basement
See photograph in Fig. 1.13, Chapter 1

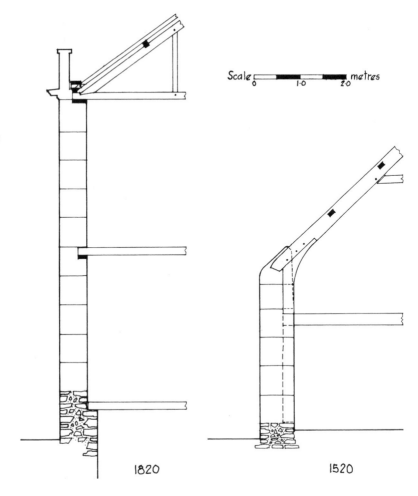

Scale 0 1·0 2·0 metres

1820

1520

been stripped off and replaced with interlocking concrete tiles (weight around 50kg/m^2) then the roof structure may have been either drastically altered or entirely re-built in order to provide a level base for the tiles. In the case of many agricultural buildings, worn and decayed thatch will have been replaced with corrugated iron sheeting (weight 14 to 15kg/m^2). The usefulness of this material, popularly known as 'curly tin', for emergency, holding repairs (making buildings safe) is discussed in Chapter 7.

Much more drastic and potentially damaging alterations are those in which the eaves level has been raised in order to enlarge and improve first-floor living accommodation. Many early cob buildings were 'improved' in this way during the late nineteenth and early twentieth centuries because their low walls, with very small first-floor windows tucked up under the projecting eaves of thatched roofs, failed to provide sufficient light and ventilation to satisfy contemporary needs. In most cases the original roof structure was entirely dismantled, though jointed cruck trusses, being difficult to remove, were sometimes left in place. In order to gain even more light and space, dormer windows were often inserted and fully hipped roofs were changed to ones of gabled type. Almost invariably, these alterations were carried out using fired bricks and softwood timber. The tops of the existing earth walls were simply levelled off and 13- or 18-inch (337 or 450mm) brickwork, built up to the required height and surmounted by a timber wall plate, provided a level base for the new roof. Potential problems that might arise from this type of alteration are: (1) additional imposed loads bearing upon walls that may already be weakened, particularly in the case of built-up gable end walls; (2) additional horizontal thrust from shallower pitched roofs bearing on the outer face of built-up lateral walls; and (3) reduced overhang at eaves and verges when thatch has been replaced with slates or tiles. Similar problems can arise from the tendency, in recent years, to convert the lofts of two-storey houses into living accommodation. Strengthening floors and roof timbers, and the insertion of internal partitions, can result in a significant increase in the dead and live loads bearing on the earth walls.

Alterations to walls

In early cob and mudwall buildings window openings were generally quite small and located, wherever possible, in south-facing walls. Walls facing north usually had few if any openings other than a doorway, and gable walls were also mostly left blank. The main oak cross-beams supporting the first floor, usually spaced at 2.0- to 2.5-metre centres, which were bedded in, and sup-ported by, the earth walls, were often located near windows and, in some cases, actually supported on window lintels. (This is not as bad as it sounds because, in buildings of this period – up to about 1650 – windows were set in substantial oak frames with equally substantial oak mullions, quite able to transfer and spread the loads imposed by heavy beams.)

It is generally accepted that, in any form of mass earth construc-tion, window openings should never be located at or near corners, because in monolithic earth walls these form the most vital structural elements. This is particularly true of walls constructed using the wet piled method where, as previously noted, non-reinforced internal corners can form points of poten-tial weakness. When mass-produced plate glass became widely available, from the 1850s onwards, it became possible to enlarge and update the windows of old cottages, and to insert new ones at little cost. Widening windows was normally achieved by inserting new lintels, cutting back the earth wall to form the new, wider opening and then building new jambs, or reveals, in fired brick, which also acted as a pillar, or pillars, to support the lintels. This work could be highly disruptive and often resulted in new cracks opening up around and above the windows, thus weakening the walls, especially in cases where the window was located close to the end of a main cross-beam or the foot of a principal rafter.

The use of inappropriate materials and ineffective repair methods

External interventions

Experience in south-west England has shown that, where repairs and alterations have been carried out to cob buildings, the materials employed were invariably stone, fired bricks or, in more recent times, concrete blocks. Earth material seems hardly, if ever, to have been used for repairs, at least during the last hundred years or so. The main reasons for this are, firstly, that the shrinkage of wet placed cob has always been recognised as a problem in bonding new work to old and, secondly, that it was not only considered more expedient to carry out repairs using masonry, but also assumed, quite wrongly, that a stronger, more permanent repair would be achieved. Unfortunately, it has to be said that many of these often clumsy and inept repairs have resulted in the development of major structural problems and occasionally actual collapse, for reasons explained below.

In the type of building with which this book is primarily concerned, the earth walls form an integral part of the structure and are fully load-bearing. Because the tensile strength of earth, even when fibre-reinforced as in the case of cob or mudwall buildings, is relatively low, any significant movement in either the roof structure or the masonry plinth is likely to result in the development of vertical cracks and fissures in the walls. When full-height cracks develop in a mass earth wall, what was formerly a homogeneous, monolithic structure becomes separated into several massive, more or less free-standing slabs of material, each weighing many tonnes. An effective repair is one that provides an efficient mechanical joint between detached sections of wall, thus stabilising the structure, and which also allows for continued minor movement within the wall itself. It is also important that the repair should not impede the free movement of moisture within the wall and that it should have physical characteristics entirely compatible with those of the existing earth material. Clearly, these requirements cannot be met by concrete, fired brick or stone set in a cement/sand mortar, as all these materials are too rigid and inflexible, and are, in addition, largely impermeable to moisture.

In exposed, unrendered walls, or those that have been simply coated with limewash, evidence of serious decay and damage and of past repairs may be clearly seen, as for example in barns, warehouses and industrial buildings, where the roof structure may also be visible. However, in domestic buildings, most of which will have rendered and plastered walls, the full extent of damage is often not revealed until the wall coverings have been removed. Although surface cracking in renders and plasters can provide evidence of possible damage and structural movement in the wall, they may be superficial and, in any event, do not always reflect the true nature and condition of the earth wall they are intended to protect (see Fig. 6.3). Full-height vertical cracks seem usually to have been dealt with in one of two ways: those located at corners were most usually patched up with fired bricks bedded in lime or cement mortar, while those located elsewhere were often simply plugged with pieces of brick and stone, and copious quantities of mortar. Work of this type may, at best, be regarded as a 'cosmetic' temporary expedient rather than a permanent repair as they fail entirely to stabilise the damaged wall and, as we shall see, usually cause more problems than they solve.

Collapsed sections of wall and deep cavities were also re-built with stone or brick, which was rarely effectively keyed into the existing earth wall, and south- and west-facing walls which had suffered from severe erosion were often 'made good' with a stone or half brick facing, also usually not bonded into the earth wall. As if all this were not bad enough, over the past 40 to 50 years it has become common practice to 'protect' earth walls by the application of hard, inflexible and largely impermeable cement/sand renders which, because they have little or no adhesion to earth substrates, need to be supported on an expanded metal or chicken wire mesh fixed to the wall with galvanised nails. Some cases have been observed where external sand/cement renders applied to metal lathing or chicken wire had been acting as a form of 'bandage', serving a similar purpose to a plaster cast applied to a broken limb. In other words, the fabric of a wall, which may actually be in need of structural repair, possibly at or near the point of failure, is contained within a strong, rigid envelope in order to keep it standing for a temporary period. Unfortunately, this is an expedient often adopted by inexperienced builders, whose clients remain blissfully unaware of the real state of their property until disaster strikes, and serious structural movement or partial collapse occurs (see Fig. 6.4).

Experience in the south-west of England would suggest that between 80 and 90 per cent of cob and mudwall houses in Britain have had their external walls wholly or partly rendered with cement/sand mortars, and it is now generally accepted that a clear correlation exists between low-permeability wall finishes and propensity to failure in earth walls. Although there is no statistical data to support this, experience would seem to suggest that serious failures in domestic earth-walled buildings have become more frequent during the past 20 years, mainly it is thought as a result of the long-term effects of cement-based renders and plasters, and low-permeability paint finishes (Keefe, 1998).

Replacement of traditional lime-based wall finishes with cement- and gypsum-based materials became common during the 1960s, when improvement grants under the various Housing Acts became widely available.

6.3
Above, removal of render coat from the gable wall of a cob house reveals the extent of repairs, alterations and deterioration. Below, detail showing failed corner repair and inserted window opening
Photos: Chris Shapland

6.4
Serious collapse of a late seventeenth-century former farmhouse following underpinning, buttressing and removal of render and plaster coats

It is possible that, in many cases, the damaging effects of these wall finishes only became apparent, and manifested themselves, after a considerable number of years. In other words, build-up of moisture in earth walls to critical levels as a result of water penetration through surface cracks may be a slow process. It is possible to monitor moisture levels in standing buildings (techniques using electronic 'remote-sensing' devices have been developed) and this could prove an interesting and potentially useful area for further research.

A wall may be regarded as outward leaning, but not dangerously so, when its vertical axis has rotated (tipped over) by more than 1.5 to 2.0 degrees (depending on its slenderness ratio). The verticality, or otherwise, of a wall can be established by means of careful measurement, including the suspension of plumb lines externally and if necessary internally, and then drawing up a cross-section through the wall to a suitable scale. It is worth carefully checking internally as well as externally for signs of movement because it is sometimes the case that what might appear to be an outward-leaning wall, especially in cob and mudwall buildings, may have formed part of the original construction rather than being the result of later structural movement. If in any doubt, a structural engineer with experience of earth-walled buildings should be employed. In the past it was customary to stabilise outward-leaning walls by means of either buttresses or tie rods, the latter intended to provide restraint to lateral walls. Clearly, the presence of either buttresses or tie rods in a building provides evidence of past structural problems, which may or may not have been satisfactorily resolved. Buttressing is an effective if unsightly way of stabilising an outward-leaning wall, but only if it has been properly designed and constructed.

It is all too common to find buttresses that have been built on inadequate foundations, in which case they may have moved out from the wall they are intended to support, thus becoming totally ineffective or, even worse, where they have been tied into the wall and then moved, pulling the structure further out of alignment. Cross-ties, sometimes of timber but most usually of mild steel, are a common sight on many old buildings and can be identified by the projecting threaded rods and nuts, and circular plates or crossed metal bars at either end. Tie rods can effectively restrain further movement in outward-leaning walls, but only if the spreader plates at either end are of sufficient size (surface area) to counteract the tensile forces of the rods acting on the earth wall which, if unchecked, can cause shear cracks to develop around the end plates.

Occasionally one may encounter an earth wall that has been stabilised by means of underpinning. This massively expensive and highly disruptive procedure is intended to deal with walls where serious structural movement has occurred as a result of supposed differential ground settlement, either below or immediately adjacent to the masonry plinth. Usually there would be no visible or obvious signs of this work having been carried out, assuming it had been completed successfully. However, there have been some cases noted in recent years of earth walls, usually of cob construction, having been underpinned and then, soon afterwards, having suffered serious structural failure or collapse (more on this issue in Chapter 7). It has become common practice to install injected chemical damp-proof courses in any building, usually those over 100 years old, which lacks a physical dpc. Mostly this is done at the insistence of mortgage lenders, presumably on the basis that even if there is no evidence of rising dampness in a house, this is, nevertheless, quite likely to occur at some point in the future. Injected chemical dpcs seem to work reasonably well in walls constructed of fired brick but are very much less effective in thick random stone walls, the cores of which may be full of large cavities. Occasionally these injected systems have been inserted directly into cob walls rather than into the masonry plinth: a recipe for potential disaster, as the introduction of large quantities of fluid at the base of a wall could well result in structural collapse. This topic is also discussed in greater detail in Chapter 7.

Internal alterations

Still on the subject of dampness in earth walls, the causes and effects of which are discussed in detail below, many of the problems that arise in domestic buildings may be directly attributed to misguided attempts at 'damp-proofing' using inappropriate modern building materials and methods. When dealing with actual or perceived problems of dampness in traditional earth buildings one needs to realise that the ways in which they are affected by, and respond to, changing environmental conditions are entirely different from those of modern buildings. According to SPAB (Society for the Protection of Ancient Buildings):

> Modern buildings tend to rely on an impervious outer layer or a system of [vapour] barriers to prevent moisture penetrating the walls, whereas buildings constructed before the mid nineteenth century generally rely on allowing the moisture which has been absorbed by the fabric to evaporate from the surface. The thickness of the wall alone may have been relied on to achieve acceptably dry conditions internally.
>
> (Hughes, 1986)

In contrast to modern buildings, traditional cob and mudwall houses with thatched roofs were constructed by local people from natural materials derived directly from the surrounding land and landscape. They are therefore in a very real sense 'organic' and like any living organism must, in order to remain healthy, be able to 'breathe' and to respond readily to variations in temperature and humidity. They need, in other words, to be in a state of equilibrium with their environment while at the same time providing healthy and comfortable living conditions for those that inhabit them. In seeking to bring traditional earth buildings up to contemporary standards by 'improvement' or alteration, great care must be taken to avoid any works that could upset this balance with the environment, as damage to the building may well be the consequence.

In traditional 'breathable' forms of construction, including earth-walled buildings, any measure that inhibits the evaporation of excess moisture from the walls and the free movement of water vapour within the building envelope is potentially damaging. The application of low permeability cement-based renders to external walls has been referred to above. In cases where this work has been carried out but the internal wall surfaces have retained their original permeable lime plaster finishes, rising or penetrating dampness, being unable to evaporate to the outside air, will migrate to the interior wall face. When this happens, the house owner is usually advised that damp-proofing measures are required to cure the problem. If an injected chemical dpc is inserted in an earth wall, the old lime plaster will be stripped off and replaced with either a strong, waterproof cement/sand plaster or some form of dry lining incorporating a vapour barrier.

If the building is undergoing major 'modernisation' then no doubt the original ground floor, of stone or brick paving, cobbles or compacted lime-ash, will have been removed and replaced with a cement screed laid on a damp-proof membrane, one effect of which will have been to transfer any groundwater, which would have formerly evaporated through the floor, into the earth walls. Other alterations that might be included in refurbishment schemes would be the blocking of redundant chimney flues, the installation of double-glazed windows and doors, and the application of low permeability acrylic resin paints and vinyl wallpapers. The combined effect of all these alterations will be not only to radically change the internal environment of the house but also to greatly increase the risk of structural failure in the event of subsequent rain- or groundwater ingress.

However, having said all this, it seems likely that, of the thousands of earth-walled buildings which have been altered in this way, only a relatively small proportion may be seriously at risk. Provided that excess moisture is prevented from entering their walls, such houses may remain unaffected and therefore serviceable for many years, but only if regular monitoring of their condition, together with careful maintenance, is carried out.

The need for maintenance

There is no doubt that earth-walled buildings, in comparison with those of conventional construction, require regular, more frequent maintenance, something which the average home buyer of today is reluctant to accept. Any building will deteriorate if it is not properly maintained, but earth-walled houses are rather more susceptible to neglect than most others, apart perhaps from those of timber-framed construction. The building element most likely to suffer neglect is the roof. Thatched and slated roofs in particular need to be regularly inspected, as rainwater penetration from above can lead to areas of the wall becoming saturated to the point where structural failure may occur. Excess moisture can also affect beam ends and rafter feet where they are bedded in the earth wall by encouraging rapid damp decay or dry rot to develop in the timber.

A sound, well-constructed covering of thatch, in addition to being aesthetically pleasing, provides both excellent protection from the weather and good thermal insulation. Only the top 50 to 75mm layer of thatch will become saturated following prolonged, heavy rain, but this moisture will evaporate rapidly in dry, windy conditions. Areas of a thatched roof most prone to decay are the ridge, which is very exposed to all directions, just above the eaves, where the roof pitch tends to be slacker, around chimney flashings, and anywhere where valleys have been formed. It is good practice, when re-thatching, to remove all old material apart from the very bottom layer, just above the rafters. Unfortunately, this is not always done and thatched roofs up to three feet (0.9 metres) thick or even more have been noted, which will have had the effect of significantly increasing the load bearing on the earth walls. Slate roofs should also be carefully examined for loose or missing slates, decayed or split lead flashings in hips, valleys, parapet gutters and around chimney stacks. Rainwater goods – gutters, hopper heads and down-pipes – also need to be regularly inspected. Mention has been made of the effects low permeability cement/sand renders on earth walls. Where these exist, as they frequently do, it is very important to ensure that, firstly, they remain in sound condition, free from major or even minor cracks, which may appear superficial, and, secondly, that they have not become detached from the earth substrate. It should be noted that driving rain is able to penetrate cracks as small as 10 microns (0.01mm) in width.

Physical damage from external sources

Ground settlement (subsidence)

The structural damage that can arise from differential ground settlement, particularly in clay soils, has been referred to earlier in this chapter. Ground movement is usually related to changes in the soil/water regime; for example, abnormally dry weather conditions may cause shrinkage in the ground or, conversely, water levels may be raised by changes in the natural drainage pattern, diverted water-courses or the removal of mature trees. Buildings sited at the foot of sloping ground may be particularly at risk from groundwater movement, and building, civil engineering or road repair works carried out in close proximity to earth walls can also undermine or weaken their foundations. If external ground levels have been raised above plinth or internal floor level, any excess groundwater present will enter the base of the earth wall, causing it to become permanently damp; and, if excess moisture is prevented from evaporating through the wall surfaces, then the wall will be seriously weakened, and its load-bearing capacity significantly reduced at the point where it is under maximum compressive stress (see Figs 6.5 and 6.6).

Plant and rodent damage

Plant growth at the foot of an earth wall can cause damage if roots penetrate the plinth or foundations below ground level. By far the worst problems, however, are caused by creepers, in particular ivy (*Hedera helix*), the roots of which can, if uncontrolled, cause severe damage and disruption to both the earth wall and its supporting masonry plinth. Walls most at risk are those that are either unrendered or covered with a weak or decayed lime-based mortar (see Fig. 6.7). Damage caused by rats is mostly confined to agricultural buildings, in particular those with exposed earth walls, though there have been reports of similar problems in residential buildings (Pearson, 1992). Rats are attracted to any building where a suitable source of food can be found, grain barns and haylofts being a particular target (one reason why granaries were usually raised off the ground and supported on 'staddle stones'). Rats will gain entry through cavities and other weak points in damp earth walls, excavating tunnels (known as 'rat runs') and nests, deep within the walls. Entry holes, which can be quite small and seemingly innocuous, are usually found either just above the masonry plinth or, in linhays and threshing barns, around first-floor level (see Figs 6.8 and 6.9). The total extent of damage is rarely, if

**Principal causes of
dampness in earth
walls**

Arrows show
direction of capillary
moisture
movement

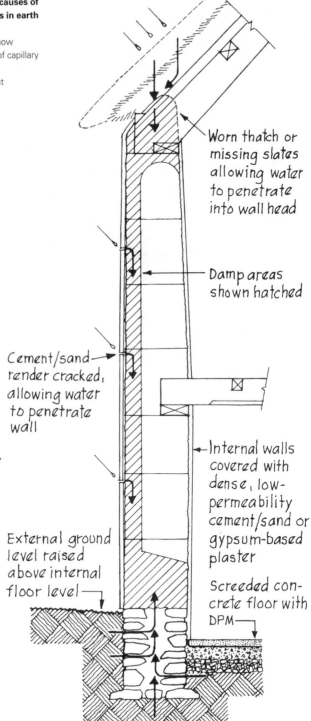

Worn thatch or
missing slates
allowing water
to penetrate
into wall head

Damp areas
shown hatched

Cement/sand
render cracked,
allowing water
to penetrate
wall

Internal walls
covered with
dense, low-
permeability
cement/sand or
gypsum-based
plaster

External ground
level raised
above internal
floor level

Screeded con-
crete floor with
DPM

6.6
Above, rising
dampness caused
by raised ground
levels on a sloping
site and, below,
detail of building
showing
structural cracking
around a window
opening

6.7
**Rampant and uncontrolled
ivy growth in a Grade I
listed cob building near
Launceston in Cornwall.
The ivy has been killed off
but not cut back**
The rebuilt wall can be seen
in Fig. 7.11, Chapter 7

6.8
Wall of a cob
building in
Somerset,
showing rat
and masonry
bee damage,
erosion and
failed cement/
sand render
coat

6.9
Collapsed section of cob wall, showing the combined effects of rat and ivy damage, and rainwater penetration through damaged roof

6.10
Former farm worker's cottage in Cornwall used as a field barn, showing severe abrasion damage caused by cattle. Also showing cob chimney breast and internal stack

ever, apparent from outside but can, if it is sufficiently widespread, undermine the wall, weakening it to the point where structural failure may occur.

Erosion, abrasion and impact damage

Surface erosion and abrasion are problems mainly encountered in the walls of unrendered earth buildings, or those where weak or decayed renders have failed and fallen away. The principal sources of erosion are wind-driven rain and, to a much lesser extent, wind-blown sand and dust. Serious erosion can be caused in several ways: (a) by 'splash-back' from heavy rain falling on hard surfaces adjacent to earth walls with low or minimal masonry plinths; (b) by erosion from wind-driven rain on walls open and exposed to the prevailing wind direction; and (c) localised erosion due to an inadequate eaves overhang or to defective rainwater disposal systems. Abrasion and impact damage, especially to farmhouses and their associated agricultural buildings, may be caused by animals, mainly cattle, and by wheeled vehicles (see Fig. 6.10). The corners of unrendered buildings are most vulnerable; also the internal walls of cow-byres where, in former times, animals would lick the walls in order, so it is claimed, to obtain mineral salts from the earth as a sort of dietary supplement. It is not unusual to see earth buildings which have been either built with rounded-off corners or corners that have been reinforced with masonry in order to prevent surface abrasion.

Dampness in earth walls and associated building materials

Moisture is always present in earth walls, as it is in most traditional building materials. This chapter is concerned with the effects of excess moisture on the behavioural characteristics, performance and durability of earth walls and, to a lesser extent, other building materials associated with them – lime- and cement-based plasters and renders, and paint finishes. Failure mechanisms associated with the presence of excess moisture in various soil types and construction methods are discussed in outline, together with the role played by low- and high-permeability finishes in controlling water vapour transmission. The measurement of moisture levels in building materials and interpretation of the data thus obtained is discussed, and some definitions given.

Moisture, for the purposes of this chapter, may be defined as that part of the total volume of a sample of soil or other material that is comprised of free water – that is, water contained within the voids, or pore spaces – which can be removed from the sample by means of oven-drying at a temperature of 105°C. Percentage moisture content of building materials may be expressed in one of two ways: wet weight or dry weight. Percentage wet weight is the difference between the wet weight of the sample – at field moisture content – and its oven-dried weight divided by its original wet weight, to give moisture weight fraction × 100. In calculating percentage dry weight, weight of moisture is divided by oven-dry weight × 100. Percentage dry weight is used in all laboratory work and for most calculations, and is used throughout the text unless otherwise stated.

Earth walls vary widely in the ways they react to the presence of excess moisture. In walls built using the wet, piled method of construction – cob and mudwall for example – cohesion is maintained only for as long as moisture in the fine soil, or binder fraction, remains below a certain level. The same is also largely true for walls constructed using adobes or clay lump. In walls built using compacted material such as rammed earth or compressed soil blocks, however, where cohesion is achieved mainly through internal friction between the individual soil particles, higher soil density and consequent low porosity mean that moisture penetration is less of a problem. The determination of what might be described as 'critical moisture content' – the point at which a soil will fail under a given load at a given moisture level – is discussed below,

Porosity/voids ratio. The moisture-holding capacity of a soil or any other building material is determined to a large extent by its porosity, which, in turn, is determined by its density. In general, the lower the density of a material the greater the number of voids it will contain. The 'term voids ratio' is defined as the ratio of the volume of voids to the volume of solids, expressed as a decimal (e.g. 0.35). Porosity has a similar meaning; it is the ratio of the volume of voids to the total volume of the soil (or other material) usually expressed as a percentage. Permeability or diffusivity is defined as the ability of a material to allow the passage of a fluid – expressed in millimetres per second (mm/s). Capillarity may be defined either as 'the ability of an unsaturated soil or rock to transmit fluid' (*Collins Dictionary of Geology*) or the 'Tendency of a liquid in a narrow tube or pore to rise or fall as a result of surface tension' (*Concise Oxford English Dictionary*).

If a particle density of $2.7g/cm^3$ is assumed (see the section on soil density in Chapter 2) except in the case of soils containing large proportions of soft shale, then it is possible to make an estimation of the porosity of a cob or subsoil sample. Reference to Chapter 2 will show that density in soils can vary from around $1,500kg/m^3$ in clayey silts and Upper Chalk up to around $2,200kg/m^3$ in pneumatically rammed gravelly sands and that the very wide variation in porosity noted is related mainly to particle size distribution and, to a lesser extent, state of compaction or packing of the soil. It follows, therefore, that the larger the fine soil fraction (grains < 425μm diameter) the greater will be the proportion of voids and the higher the porosity. The highest percentage porosities will be found in silty clays, which, because of the large specific surface areas of the clay and silt particles, will contain a significantly higher proportion of micro-pores (pores less than 50μm diameter) or capillaries. This has an effect not only on density but also, and more importantly, on the permeability and moisture diffusivity of soils.

In Chapter 2, Fig. 2.3, the various physical states of the fine soil fraction, from solid to liquid, are shown in relation to moisture content. Equilibrium moisture content (emc) in an earth wall is reached when the moisture content of the soil is equal to that of the surrounding atmosphere. At the foot of a cob wall this has been found to vary from around 2 to 5 per cent wet weight (Trotman, 1995).

The movement of water through traditional building materials (permeability or diffusivity) and their ability to absorb water or lose it through surface evaporation, is determined to a great extent – particularly in the case of soils – by their particle size distribution characteristics. In a study carried out at the University of Grenoble (Dayre and Kenmogne, 1993) using gamma ray spectrometry, two compressed earth blocks with widely differing psd characteristics were placed in sealed containers and their rates of moisture diffusivity measured. The results showed that, in a sandy clay soil (85 per cent sand, 8 per cent silt and 7 per cent clay), the rate of capillary movement was 50mm per hour, whereas in a silty clay (20 per cent sand, 66 per cent silt and 14 per cent clay) the rate was measured at 4mm per hour. The dry densities and porosities of the two blocks were 1,840kg/m^3 and 31 per cent for the sandy clay, and 1,550kg/m^3 and 40 per cent for the silty clay, thus demonstrating the importance of all these factors – particle size distribution, density and porosity – in determining the behaviour of soils affected by excess moisture. It should be noted that, although coarse-grained, sandy soils have the ability to absorb and transmit water quite rapidly, they will lose it through surface evaporation with equal speed. For example, an experimental free-standing cob wall, 450mm thick and 1.5 metres high, which was built in the grounds of Plymouth University in order to monitor moisture movement in earth walls, dried down to equilibrium moisture content in 28 days (Greer, 1996).

Wall finishes – plasters, renders and paints

Apart from soil characteristics and moisture levels, the other single and possibly most crucial factor in determining propensity to failure in earth walls is the nature of applied finishes, in particular, their permeability, and it has been accepted for some years now that low-permeability finishes can have a detrimental effect on the performance and durability of earth walls. The fundamental difference between cement- and lime- or earth-based renders and plasters is that the former are intended to provide an impermeable 'weatherproof' outer skin and the latter a 'sacrificial' coating designed to prevent erosion and abrasion of the earth substrate while allowing the free passage of water and a degree of thermal and moisture expansion to take place. Cement-based materials are brittle and lime-based materials relatively flexible. Although the types of aggregates used and their proportions are similar in both cement- and lime-based finishes, the nature of the two binders is quite different. The dry bulk density of cement is about twice that of lime – thus increasing the weight of a render coat by about 20 per cent. Unlike lime, which acts in a similar way to clay binders in soils, cement paste, when it cures, develops fibrous shoots of calcium silicate hydrate which expand to fill the pores between the aggregates, thus significantly reducing their porosity. Some laboratory tests have been carried out in order to compare the water vapour transmission (WVT) characteristics of lime-based materials with those of other products. In the two research papers referred to below, a standard dish assembly apparatus was used, so the resulting data, expressed in grams per square metre of water over a 24-hour period, are directly comparable. In the first paper (Jacob and Weiss, 1989) a 1:3 cement/sand mortar showed a WVT rate of 50.4gm^2 24h, whereas a 1:2:9 cement/lime/sand mix showed a rate of 170.4gm^2 24h, indicating that the lime-based material had a permeability three times greater than that of the cement/sand mortar. A pure lime/sand mortar was not tested but would probably yield similar, if not better, results than that of the composite cement/lime mortar. The second research paper (Trotman and Boxall, 1991) compared traditional limewashes with a proprietary acrylic resin-based exterior emulsion paint. The emulsion paint showed a WVT rate of 50.2gm^2 24h, while the five limewashes tested (which varied according to the binders used) showed values in the range 128.6 to 172.7gm^2 24h. These results demonstrated that the limewashes were significantly more permeable than the emulsion paint, again by a factor of approximately three.

Other disadvantages of cement-based renders are that they have very poor adhesion to earth substrates, even when a mechanical key is incorporated, and that they invariably develop cracks and fissures, mainly as a result of thermal movement which, even when they are very fine – down to 0.1mm wide – will allow rainwater to penetrate (BRE Digest 410, 1995). Water penetrating the wall is then trapped because the low-permeability render prevents it from evaporating. This phenomenon has been observed by the author in cob houses where, even in mid-summer, the outer parts of west- and south-facing walls have shown measured moisture levels of 17 to 20 per cent (dry weight). However, unless one also takes moisture readings internally and from the core of the wall it is not possible to assess the total extent of moisture penetration.

The measurement of dampness in earth walls, plasters and renders

The ability to measure accurately the actual moisture content of earth walls and associated materials is essential in carrying out diagnostic surveys. Three methods may be employed for field surveys, and these are: (1) on-site readings taken using a hand-held electrical conductance damp meter; (2) on- or off-site measurements taken by placing drilled samples in a calcium carbide meter; or (3) taking drilled core samples, placing them in an air-tight container and accurately measuring wet and oven-dried weights in a laboratory.

Electrical conductance damp meters have been in use by building surveyors for many years. They were designed specifically for measuring moisture in timber and may be reasonably effective in this role, but for moisture measurement in other building materials they appear to be of only limited use. At best, they may provide an indication of either a complete absence of moisture (rare in traditional buildings) or the presence of moisture, the extent of which is not accurately determined – a material may be either bone dry, relatively damp or apparently saturated. The 'Speedy' calcium carbide moisture meter is an expensive piece of equipment in which a fixed weight of soil, or other powdered/disaggregated material, is mixed with calcium carbide powder in a sealed pressure flask and then shaken. Reaction of the powder with the moisture in the soil or aggregate produces acetylene gas, creating a build-up of pressure related to the moisture content, which is displayed as a percentage on the built-in pressure gauge.

In order to determine whether it might be possible to adapt an electrical conductance meter (the 'Surveymaster' manufactured by Protimeter plc) for measuring dampness in cob buildings, some basic tests were carried out. The H_2O dial on the 'Surveymaster' is graduated from 0 to 100; 0 to 25 being regarded as a percentage and 25 to 100 as relative moisture content up to saturation. The meter's probes were inserted into two cob/soil samples and readings taken. The first (clay/silt) sample was air dry and, after some wild fluctuation, showed a reading of around 25 (per cent). Oven drying indicated a moisture content of 1.5 per cent dry weight. The second, a sandy silt cob sample, showed meter readings fluctuating between 60 and 80, whereas the oven-dried results were 3.8 and 4.1 per cent. It seems, therefore, that, if readings are taken at face value, the result is to massively overestimate the amount of moisture present. To summarise, field and laboratory tests would seem to suggest that, for the measurement of moisture content in earth walls electrical conductance meters are entirely unsuitable, that calcium carbide moisture meters are reasonably accurate (to within ± 0.5 per cent) for higher moisture contents, and that weighing and oven drying is the preferred method.

Hygroscopicity, soluble salts and freeze/thaw conditions

Soluble salts may occur naturally in soils or be present as a result of rising dampness. In earth walls these salts, which are mainly nitrates, chlorides and, to a lesser extent, sulphates, are drawn by capillary movement towards the surface layer of the wall, where, in dry conditions, they will crystallise within the pore spaces, up to 20 to 30mm from the wall face, and expand, thus forcing the soil particles apart. This cyclical process of hydration and dehydration can cause extensive spalling (breaking up of the surface layers) but mainly, it should be noted, in chalks and clayey silt soils containing numerous micro-pores. In coarse, sandy soils with an open pore structure soluble salts are less of a problem, because any pore water present is able to evaporate quite rapidly. Both nitrates and chlorides are particularly hygroscopic; that is, they have the property of absorbing water vapour from the atmosphere. The presence of these salts may therefore have the effect of significantly increasing moisture levels in both earth and masonry walls. A comprehensive condition survey would need to take into account the possible presence of soluble salts, which can be detected by adopting the survey and testing procedures described in BRE Digest 245 (1985). Increase in internal pore pressure as a result of water expanding as it heats is unlikely to be a problem in northwestern Europe, as diurnal variation of air temperature in excess of 20°C is quite unusual. On the other hand, pore pressure caused by the rapid freezing and thawing of pore water, at or just behind the wall surface, is quite common; it can cause serious disruption in earth walls, but is more likely to be a problem in micro-porous soils such as clayey silts and, in particular, chalk, which is very moisture-retentive.

Determination of critical moisture content in earth walls

It is very useful, when dealing with cob or mudwall buildings whose walls have become excessively damp, to have some idea of what the soil's critical moisture content (CMC) is likely to be. It has been noted earlier in this chapter that, in a wall covered with a strong cement/sand render applied to a metallic reinforcement, the render coat may be acting as a rigid, impermeable envelope, effectively containing an earth wall that may be at or even beyond the point of failure. In fact, several cases have been recorded where serious collapse has occurred immediately following the wholesale removal of plaster or render coats. However, this is a situation that can be avoided, or at least controlled, by following the suggested diagnostic survey procedure described in the next chapter.

In Chapter 2 the process by which compressive strength decreases with increasing moisture content in earth walls constructed using the wet, piled method has been fully described. Intact soil samples removed from earth walls immediately following their collapse have shown a moisture content as low as 7 per cent in some cases, so clearly reliance on moisture measurements taken from whole samples can be very misleading. What needs to be borne in mind is that, in standard soil consistency tests carried out in accordance with those described in BS 1377 Part 2, only the fine soil fraction (material < 425μm diameter) is used. When a sample of damp cob or clay lump is removed from a building it is not possible to separate out the fine material because of the cohesive nature of the clay fraction and the surface tension of water between individual soil grains. The plastic limit, which represents the point at which a soil will start to deform under load, is therefore only directly relevant to soils comprised entirely of fine material, which makes the prediction of critical moisture content in soils containing, as they usually do, a significant proportion of coarser particles, somewhat diffi-cult. However, if a large enough sample of damp soil can be obtained, its moisture content measured, and its particle size distribution characteristics and plastic limit established, then an estimated value for its critical moisture content may be arrived at by extrapolation.

On the basis that in most soils – other than those that contain a large proportion of porous aggregates – moisture content will increase as par-ticle size decreases, and that, in soils comprised entirely of material less than 425 microns, CMC will be at or just above the plastic limit, the following pro-cedure may be adopted:

1   Measure, by weighing and oven drying, moisture content (dry weight) of whole sample.
2   Measure percentage of material < 425µm by wet sieving and oven drying.
3   Measure plastic limit (per cent moisture content).

Example:

A sample of cob, known locally as 'witchert', from Haddenham, Buckinghamshire.

Moisture content of whole sample = 8.9%

Material < 425µm = 40% of sample

Plastic limit = 26%

Calculation: 100% (if all material were < 425µm) minus 40% = 60%; therefore, reduce percentage moisture content at the plastic limit by this amount, which gives an estimated CMC of 10.4%.

In this case an increase in moisture content of as little as 1.5 per cent would result in deformation of the soil and, eventually, structural failure. In order to test the validity of this theoretical procedure, the material was also reconsti-tuted (re-mixed, with fresh straw added) and formed into two 200 × 100mm cylinders, which were then tested for unconfined compressive strength. One specimen was air-dry (moisture content *c.* 2.0 per cent) and failed at 1,000kN/m^2, while the other, at a moisture content of 10.2 per cent, failed at 200kN/m^2, a reduction in compressive strength of 80 per cent; serious, though still just within safe limits for a two-storey wall. In this context, there-fore, CMC represents not the actual point at which failure will occur but rather a threshold, or minimum wet compressive strength, which, if exceeded, will lead to structural failure through either deformation of the soil or the devel-opment of shear planes, depending on the soil's particle size distribution and the size and composition of the clay fraction. Field survey procedures for recording and measuring dampness in earth walls are described in the fol-lowing chapter.

Chapter 7

The diagnostic survey and repair of earth buildings

A suggested diagnostic survey procedure for earth-walled buildings

The following notes and suggested diagnostic survey procedures are based largely on the author's personal experience of observing and recording earth-building failures, mainly in cob walls, over a number of years. The checklist is intended to cover all the various factors (described above in Chapter 6) that may contribute to failures in earth structures, but in outline only and should not, therefore, be considered in any way definitive.

One problem frequently encountered by building surveyors is that of identifying the material from which a building with totally rendered walls has been constructed. Even some building owners are unaware that the houses they occupy are constructed mainly of subsoil! Wall thickness is usually a good guide to identifying cob or mudwall buildings; any wall whose overall thickness is 550mm or more is likely to be made of earth. Other points to look for are undulating wall surfaces, chamfered inner window reveals, slightly rounded corners and inner walls tapering off towards eaves level. Plinth heights, which show considerable regional variation, range from as little as 300mm above ground level up to first-floor level. Clay-lump buildings are rather more difficult to identify, as are cob or mudwall buildings which have had a brick or flint facing applied to their walls.

What should be emphasised at this point is the need to positively identify the materials from which the wall is constructed; in other words, what is being removed when samples are taken – earth, stone, brick or mortar (which may be either lime- or mud-based). In order to answer this question it is first necessary to establish the height of the masonry plinth, even if this involves, as it often will, the removal of a vertical strip of plaster.

Samples are normally taken from both external and internal walls and it is desirable to remove sections of plaster and paint so that they can be examined and tested separately. Plaster coats may be either wetter or drier than the cob or stone substrate, depending on their basic constituents, and can provide useful clues concerning the behaviour of apparently damp walls, as can paint finishes.

Condition surveys of earth-walled buildings differ from those appropriate for 'conventional' traditional buildings in two main respects. Firstly, because of the nature of the raw material and its susceptibility to the effects of excess moisture, great care needs to be taken, and special investigative techniques employed, in the examination of potentially unstable walls. Secondly, the presence of structural movement or discontinuities in domestic buildings are often concealed behind render and plaster coats which may, in themselves, be forming temporary structural elements in walls suffering from incipient or actual structural failure. A suggested six-point diagnostic survey procedure is shown in outline in Table 7.1 and described in more detail in the checklist set out below.

Table 7.1 Six-point diagnostic/condition survey

Stage	Items to be recorded or surveyed
1	**External environmental factors.** Orientation, elevation, topography, exposure, vegetation/planting, including in particular any close to, or growing into the wall. Land and domestic drainage. Local soils and surface geology. Climatic conditions.
2	**Description of building.** Roof structure, roof covering, first floor structure, openings – doors and windows, masonry plinth, earth walls, chimney stacks and flues, wall finishes.
3	**Maintenance and repair history of the building.** Including any major alterations or extensions; based upon information from the owner and visual inspection.
4	**Condition of building, and description of observed failures and defects.** Roof, external and internal walls. External and internal cracks, fissures and structural movement in both earth wall and plinth. Wall finishes – external and internal, for evidence of dampness.
5	**Dampness survey.** Measurement of moisture levels in affected areas, including hygroscopicity and the presence of soluble salts.
6	**Analysis and testing of earth material from walls.** In cases where moisture levels are considered to be excessive, in order to establish, *inter alia*, critical moisture content.

Checklist: characteristics and features of building and site to be recorded

(1) External environmental conditions, including: orientation, local geology and soil classification, degree of exposure – height above sea level (Ordnance datum), contour intervals (steepness of slopes), proximity of mature trees and other buildings. For any walls particularly exposed to the south and west calculate, if required, the Driving Rain Index (described in BS 8014 1992). Land drainage, including water table (if known). Foul and rainwater drainage. Check external ground levels relative to floor levels. Nature of surfaces at base of walls, including any stored or stacked material and vegetation – mature shrubs and climbing plants. Note weather conditions at time of survey and for the preceding month.

(2) Details of original construction that may have contributed to the structural movement/failure, excluding characteristics of earth wall material (but refer to item (6)). Roof structure; record constructional details and any movement that may have occurred (see also item (4)). Type of roof covering – if thatch, try to measure its depth. Structure at first-floor level; joists and beams, also staircases. Note window details, in particular lintels, in areas of failure. Height of stone plinth, if visible and if one exists. Thickness and height of walls – calculate slenderness ratio, taking into account any tapering-off of upper wall. Chimney stacks and flues, noting particularly any that are blocked or truncated.

(3) Later alterations, repairs or inappropriate maintenance that may have contributed to the failure. In particular look for alterations that may have increased the loading on the walls or produced eccentric loading at the wall head; also new openings in the walls adjacent to corners, beam ends or truss feet etc., any of which may have weakened or over-stressed the earth wall. A calculation of stresses at key points can then, if required, be related to the measured compressive strength of earth samples and estimated critical moisture levels. Look also at the following: types of renders, plasters and finishes, both internal and external. The presence of buttresses and wall ties, indicating earlier failures, and how useful or damaging such measures appear to have been; any other repairs (if visible) that may have been carried out using masonry (stone, brick or concrete blockwork) or other non-earth materials.

(4) Condition of building. Detailed survey work should be restricted to those parts of the fabric where deterioration or damage, through either moisture penetration or structural movement, may have contributed to the observed failure. This is because some items that would normally be included in a standard house condition survey may not be relevant; for example, electrical wiring, heating and plumbing (other than more obvious, serious internal leaks), condition of floors and internal decorations. The following items, where relevant, would need to be included:

Roof (external): check condition of roof covering, normally slate or thatch in domestic buildings, looking particularly at areas where rain may have penetrated the roof. Ridge and hip tiles, ridging (thatch), valley or parapet gutters, lead soakers (where slates are mitred at hips), chimney flashings/fillets (especially in thatched roofs). Check eaves overhang, also condition of gutters and hopper-heads, either of which may be blocked. Check condition of chimney stacks, looking for any signs of leaning, settlement or fissuring, or leakage in flaunchings around chimney pots.

Roof (internal): look at general configuration of roof timbers, noting any evidence of failure or movement – lack of lateral or longitudinal restraint (failure, or absence of, collar braces, tie beams or hip rafters), or racking, which can be a particular problem in fully hipped roofs, inappropriate repairs or alterations. In thatched roofs, measure or estimate thickness of material at eaves and ridge, and a point in between (in order to check roof loads if required).

Walls (external): in addition to the items included in item (3) above, the following should be noted: verticality. This should be checked both internally and externally, because it is often the case that what appear to be outward-leaning walls may have formed part of the original construction rather than representing later structural movement. Cracks and fissures: most cracks in earth walls are vertical or diagonal rather than horizontal. However, whereas in most cases a wall is rendered or plastered, horizontal cracking may appear as a result of settlement in either the wall itself, the masonry plinth or, far less common, the ground underlying the building. In these cases horizontal cracking and outward displacement of parts of a detached render coat may be indicative of structural movement resulting from a damp earth wall having slumped. Vertical cracking in window or door reveals may occur as a result of localised shear failure. Note that the disposition of render cracks may not always reflect the actual position or extent of fissuring in the earth wall beneath. Therefore, in cases where serious surface cracking has occurred and the render coat has become detached from the earth substrate, sections of the rendering will need to be removed, together with internal plastering, in order to assess the extent and severity of any structural cracks in the earth wall. Care should be taken when removing material, especially from the base of the wall, in cases where preliminary investigation has revealed the presence of high levels of moisture. The internal and external wall surfaces should be examined for evidence of dampness. Check rainwater down-pipes and drains, and look for other possible sources of rising or penetrating dampness. The lintels, and their bearings, above all openings should, if possible, be checked for decay, damage or movement. Any cracks, bulges or movement in the masonry plinth should be investigated for possible failure (internal shearing or loss of bond).

Walls (internal): check type of wall finish and look for signs of rising or penetrating damp; also tanking and/or dry-lining, which may be concealing evidence of moisture in the earth wall and possibly exacerbating its harmful effects. Check also for evidence of structural movement, in particular at corners, junctions with ceilings and cross-walls, and where joists, beams and exposed truss feet enter the wall. Look also for gaps between floorboards and walls.

(5) Suggested procedure for dampness surveys. In those parts of the walls where moisture concentrations are thought to be highest, proceed as follows:

- Establish plinth level (where one exists) by careful removal of a 100 to 150mm wide vertical section of the external render or plaster coat.
- From the section of cob wall thus exposed carefully remove a 2 to 3kg sample of material. Place immediately in an airtight container for weighing, oven-drying, re-weighing and subsequent analysis. If necessary, fill cavity with rammed cob or a cob block-bedded in either lime/sand or mud mortar.
- In cases of rising damp, take a vertical line above the disturbed area, up to first-floor level initially, higher if necessary. At 500mm intervals externally, cut out 100mm square sections of render and remove about 100 grams of cob, placing the samples immediately in an airtight container. Ideally, the same procedure should be followed internally. However, in order to avoid excessive disruption a 9 to 20mm diameter masonry drill bit 200 to 300mm in length may be utilised, though clearly this would result in a smaller sample. Moisture content may be either measured on-site using a calcium carbide moisture meter or in a laboratory. Part of each sample may, if required, be kept back in order to measure its hygroscopic moisture content, in accordance with the test described in BRE Digest 245. For penetrating damp higher up the wall or just below eaves level a similar procedure may be adopted; however, in these cases one would tend to work down from, and to either side of, the source of dampness rather than up from plinth level.
- Carry out analyses of the samples obtained as quickly as possible in order to establish whether moisture levels in the wall are above

or below the critical moisture content determined by the results of the laboratory tests referred to below in item (6).

- By working through the items included in the comprehensive check lists referred to above and following the procedures outlined in the dampness survey, a complete picture of the symptoms and mode of failure can be built up, enabling appropriate remedial works to be specified.

- In buildings which have suffered serious collapse, all that can be done is to measure moisture levels in samples taken from as near as possible to what appears to be the point where the failure originated – this to be done as soon after the collapse as possible in order to obtain accurate readings. The samples to be analysed and tested as above, and the site of the collapse to be carefully observed, measured and recorded so that the cause(s) of failure may be determined.

(6) Analysis and testing of earth samples. In addition to the survey procedures referred to above, which are designed to measure and record the extent and degree of moisture penetration, the following tests should be carried out by a specialist soils laboratory, on the 2 to 3kg sample removed from the failed wall. Note that, although some of these tests could be carried out in the field – that is, outside a laboratory – interpretation of test results requires specialist knowledge of, and experience of working with, soils used for building.

- Simplified particle-size distribution analysis by dry-sieving (coarse aggregates over 2.0mm diameter) and wet-sieving (particles between 2.0 and 0.063mm). Sample size either 300g for fine soils or 600g for those containing a large proportion of coarse aggregates.

- Separation and weighing of any significant amounts of organic material.

- Analysis of fines (material less than 63μm diameter) either by sedimentation – using pipette or hydrometer – or by means of a low-angle laser light-scattering instrument.

- Determination of plastic and liquid limits (and therefore of plasticity index).

- Measurement of linear shrinkage, if required.

- Calculation of dry density.

- Estimation of porosity based upon measured dry density.

- If a large enough sample can be obtained (4.5 to 5.0kg) then, in addition to the tests outlined above, the material can be reconstituted, formed into a cylinder and subjected to uni-axial, unconfined compressive loading in order to obtain comparative data on its compressive strength.

By analysing the results of the above tests it should be possible to compile a soil and moisture profile of the earth wall, which would enable the performance of the material, including its probable critical moisture content (point of failure) to be predicted.

The repair of earth buildings

Principles of repair

The philosophy of 'conservative repair' or 'minimum intervention' advocated for over a century by SPAB (Society for the Protection of Ancient Buildings) is particularly relevant in the case of repairs to earth buildings. The aim, especially in the case of historically important structures, should always be to repair and consolidate, without causing unnecessary physical disruption, in order to ensure long-term stability and optimum performance, and some guidance as to how this may be achieved is offered in the remaining pages of this chapter.

However, it must be emphasised that, although the advice provided is based upon various repair works carried out successfully over a 15-year period, nevertheless, practical experience in the field of earth-wall repair is fairly limited, which is why general rather than specific advice is offered.

Contrary to popular belief, mass earth is an inherently stable material, provided it is kept dry, and will tolerate a great deal of abuse. Because cob walls are at least 450 to 600mm (18 to 24 inches) thick, surface erosion is not normally a serious problem. An additional advantage of such thick walls is that 'stitching' across cracks and fissures to achieve an efficient mechanical joint is relatively easy, as is the reinstatement of the cavities and hollows that can result from more serious erosion, abrasion and minor damage. Attempts should never be made to realign outward-leaning sections of a cob wall by the use of hydraulic jacking or levering devices. Walls must always be stabilised or consolidated as they stand, except in cases where an unacceptable degree of outward lean has developed. Where outward lean exceeds the safe limit, and is combined with severe cracking, careful dismantling and rebuilding may be the only answer.

Walls constructed of clay lump are, of course, quite different from mass earth walls. Essentially, clay lump is a form of small-unit construction, not unlike stone or brick walling, and walls are, correspondingly, more slender, usually 225 to 300mm (9 to 12 inches). One would assume that, as in any form of conventional masonry construction, the mortar joints would form the weakest points in the wall. However, if the clay lumps have been laid in a mud mortar rather than a weak lime/sand mix, then very strong adhesion would exist between the two elements, though the tensile strength of the clay lumps would be greater than that of the mortar because of the straw binder they contain. Movement in a clay-lump wall could, nevertheless, result in serious fracturing, and severe localised erosion could also pose a problem in walls that are relatively slender. Repairs to decayed or damaged clay-lump walls are normally carried out by carefully removing sections of damaged wall, lump by lump, and their replacement with new blocks of similar dimensions, fabricated from either salvaged material or new earth and straw obtained from nearby. Badly eroded areas could be faced up with half or one-third width blocks keyed into the existing wall. Smaller, brick-sized clay/straw blocks, with reinforced mortar joints, could also be used to stitch across structural cracks. Both these repair techniques are described in detail below.

Appropriate repair materials

Repairing a mass earth or clay-lump wall with stone, brick or concrete blocks bedded in a cement/sand mortar may be compared to mending an old, worn garment with a strong leather patch. The long-term effects are likely to be equally disastrous because in many respects these hard, inflexible materials are incompatible with earth walls. It has been noted that, in order to perform efficiently, earth walls need to be able to shed excess moisture freely. One effect of inserting hard, relatively impermeable materials into earth walls is to impede the free movement of water vapour, thus concentrating excess moisture at the interface between the two materials. Structural failure could well occur at this point because of the build-up of moisture levels, especially in cases where there is a poor or non-effective bond between the inserted masonry and the earth walling, and where a strong, impermeable cement-based render has also been applied. Earth walls which have, in the past, been repaired or partly rebuilt using stone, brick or concrete blockwork are, there-fore, much more likely to fail when they become damp or are subjected to excessive or eccentric loads.

Hardwood timber is probably the only structural material that can be regarded as being compatible with mass earth because it is both flexible and vapour-permeable, while having a greater strength-to-weight ratio than both mild steel and reinforced concrete. In addition to these qualities, oak in particular is very durable and resistant to both damp decay and fire damage. For certain types of structural repair the use of timber is obviously essential, and its advantages as a substitute for concrete and steel for structural repairs to earth buildings are not perhaps sufficiently recognised.

Another material that has, in recent years, been quite widely used in repairs to earth walls is stainless steel, in the form of (1) expanded metal lathing (Expamet Building Products) and (2) less well known but equally

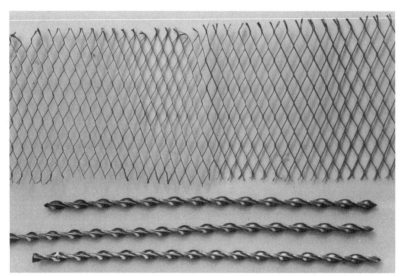

7.1
The reinforcers.
Above, expanded
metal lathing
(eml); below,
Helifix wall ties

useful, helical wall-ties, manufactured and marketed by Helifix Ltd (see Fig. 7.1). These products are used mainly to reinforce joints in structural repair works but do have other non-structural applications. There are some who might argue that such 'high-tech' materials are inappropriate for use with what is essentially a 'low-tech', even primitive and archaic, form of construction. However, their use often enables effective repairs to be carried out in ways that are relatively non-invasive and certainly less physically disruptive than conventional solutions, which often involve large-scale demolition followed by rebuilding with the inappropriate materials referred to above. Both products are corrosion resistant, have high tensile strength and are sufficiently flexible to respond to bidirectional movement in earth walls without fracturing. Although their use may not always be strictly necessary in conservation terms, the incorporation of such metallic reinforcement tends to be favoured by architects and structural engineers, who may otherwise be somewhat sceptical about the use of raw earth as a repair material. Lime stabilisation of soil for earth plasters has been discussed in some detail in Chapter 4. Stabilisation with hydraulic lime may be justified in certain cases, in particular for emergency repairs carried out in unfavourable climatic conditions or where access is difficult and repairs need to be carried out quite rapidly. These issues are further discussed in the following pages.

Clearly, the most appropriate way to maintain the structural and historical integrity of earth-walled buildings is to carry out repairs using the original raw materials, subsoil and, where necessary, some form of organic fibre reinforcement. Until quite recently, however, problems associated with the drying-out shrinkage of wet earth and the bonding of new work to old have acted as a deterrent to the use of earth for repair and reconstruction.

In fact, recent experience has shown that these problems can, to a great extent, be overcome and some suggested repair methods, using earth in various forms, are described below.

Making buildings safe, specifying repairs and monitoring dampness

In cases where a diagnostic survey has revealed evidence of either structural movement or excessively high levels of dampness, possibly both, those parts of the building considered to be most at risk will need to be stabilised or otherwise made safe, and further, more detailed survey work carried out. Among the most serious problems in earth walls are those where either the masonry plinth or the base of the earth wall show clear signs of bulging, cracking or other evidence of movement, and where this is accompanied by high levels of moisture. When actual or potentially serious structural problems have been identified, the wall or walls in question will need to be made safe by propping and shoring in order to ensure that (1) the building, or at least parts of it, can continue in occupation, (2) the general public are not put at risk and (3) further investigative work and, if necessary, emergency holding repairs can be carried out in safety. The design and supervision of the type of stabilisation works shown in Fig. 7.2 should be carried out only by a qualified and experienced structural engineer.

By excavating 300 to 400mm square trial pits at, say, two-metre intervals along the foot of the wall, one can establish the depth and condition of the masonry plinth and foundation, if one exists, and the extent to which water may be entering the wall from below. Standing water, for example, would indicate a high water table, which would require the installation of a land drain. One might also be able to observe and assess the condition of underlying soils or rock strata. In many buildings a cement/sand render coat will extend down to, or just below, external ground level, in which case vertical slots, around 150mm wide and extending at least 1.5 metres above ground level, will need to be cut out with a disc cutter, in order, firstly, to establish plinth height and condition and, secondly, to measure the vertical extent and seriousness of damp penetration in the earth wall above. When vertical slots are cut into the render coat in this way, the risk of collapse that might result from large-scale, indiscriminate removal of the render coat is much reduced. Controlled dismantling of a failed wall is clearly preferable to the major damage that would result from a sudden collapse. In walls that are very damp but have not yet developed internal shear cracks, the vertical slots will enable the wall to start drying out and allow regular monitoring of moisture levels to be carried out in safety.

It has been suggested that large amounts of free water in an earth wall may lead to the all-important clay fraction being leached out. However,

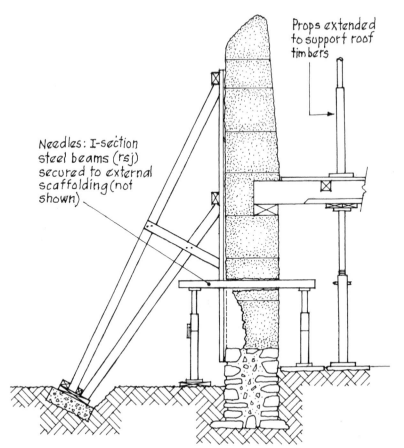

Props extended to support roof timbers

Needles: I-section steel beams (rsj) secured to external scaffolding (not shown)

this does not seem to be the case because, although some dispersion and realignment of clay particles may occur, inter-particle cohesion will be re-established, albeit in a somewhat altered form, when the material has dried out. Irreversible damage occurs only when the material is unconfined (which is not the case when a hard, metal mesh reinforced, cement-based render has been applied) and contains sufficient free water to induce failure under a given load. When, in other words, its critical moisture content has been exceeded.

In cases where water penetration has occurred higher up the wall or through the roof and into the wall head, leaving the base of the wall unaffected, some short-term remedial or preventive works may still need to be carried out. Water penetration into the wall head is a particular problem in earlier buildings where the principal rafters were often supported on timber bearer plates bedded directly into the earth wall rather than being carried on a continuous wall plate or ring beam. When the earth reaches a plastic state and starts to deform under load, the roof timbers will begin to settle, causing the affected section of wall to sink and tilt outwards, taking the rafters with it. In later build-ings with fully triangulated main roof trusses supported on wall plates, very damp sections of wall will sometimes slump and fall out under their own weight, leaving the roof structure more or less intact. Clearly, when moisture is present in the wall head to this extent, temporary holding repairs need to be carried out without delay; the main priorities in this instance being (1) to remove the source of penetrating dampness, (2) to restrain further movement in the roof structure and (3) to allow the damp wall to start drying out.

It is a fairly simple matter to carry out repairs to slated or tiled roofs, unless of course the roof structure itself has moved. Worn or decayed, saturated thatch, on the other hand, may pose more of a problem. An experi-enced thatching contractor would need to be consulted for advice concerning the possibility of temporary repairs. If decay is very far advanced and total re-thatching is required, the old thatch can be removed and the roof temporarily covered with corrugated iron sheeting (this material has the advantage of pro-viding a lightweight covering that will stiffen and stabilise the structure, and help restrain further movement in the rafters). The propping arrangement shown in Fig. 7.2 would need to be extended above first-floor level in order to support the roof structure, and cross-ties installed at eaves level to prevent further outward movement of the rafter feet. When the roof has been sta-bilised, the render coat may be carefully removed from parts of the damp wall to allow it to start drying out. It should be noted that there are various ways of stabilising old buildings and making them safe, not all of which are discussed in detail here as they are fully described in other publications.

The repair of hollows, cavities, minor surface erosion and abrasion

Although this type of minor damage is to be expected in exposed earth walls, the removal of hard, cement-based render coats will often expose similar problems. For example, the wall surface may have been eroded by water running down behind a detached render coat and, if the render had been applied to a steel or wire mesh nailed to the wall, its removal may cause weak or damaged sections of earth to break away. Hollows and cavities less than 100mm deep may be filled with a soil/straw mix similar to that of the existing cob wall but with the coarser aggregates, more than 10mm diame-ter, screened out. In all cases it will be necessary to cut back the existing cob to form a square cavity ready to receive the new material. A flat, level base is essential, and the tops and sides of the cavity should be squared off and, if possible, undercut. In order to further secure the repair, an effective mechanical key between existing and new material can, if required, be achieved by using shards of slate, galvanised 100mm slab nails or non-ferrous metal connectors, such as Helifix wall ties, driven into the thoroughly pre-wetted cob at an angle of about 20 degrees above the horizontal. The damp (not wet) repair material can be hammered or rammed into the cavity in vertical layers about 25 to 30mm deep, allowing a few hours or up to a day according to climatic conditions, for each layer to start drying out before adding the next. The final layer should project forward of the wall face to allow for (1) periodic knocking back to control shrinkage and (2) paring down when the repair material has completely dried out (see Fig. 7.9).

In cases where the entire surface of, for example, a south or west-facing wall is eroded but only superficially, no more than 25 or 30mm back from the original wall face, a lime-stabilised earth mortar, as described in Chapter 4, could be applied. Having first repaired any hollows or cavities, as described above, the earth mortar – containing lightweight aggregates if required – would be applied in one or two coats, depending on the method to be adopted. As with lime plasters, the adhesion of an earth plaster to an earth substrate is improved if it is thrown, rather than floated or trowelled, on to a rough (pitted and stony) wall surface; the only preparation then required is to brush off any dust and loose particles, then to thoroughly dampen the surface by repeated use of a fine spray. If the first coat were hand-thrown or rough cast, this would provide a sufficient key for a second, floated coat, otherwise the first coat would need to be scored to provide a key for the second. A blown or pumped mortar, on the other hand, would probably be applied in one operation, with the surface being re-worked or knocked back as necessary as it dries out (this type of mortar would probably need to be applied very wet).

The repair of structural cracks and fissures

When what was formerly a mass, monolithic wall has, as a result of structural movement, separated into two or more discrete, more or less free-standing elements, the only way it can be effectively stabilised is by means of crack stitching. The principal aim of this non-disruptive repair technique, which was developed by the SPAB at the beginning of the twentieth century for use in churches and other historic masonry buildings, was, and still is, to retain as much of the original building fabric as possible. Repair rather than restoration. A structural crack is one that extends through the whole depth of a wall from the outside to the inside face and which has developed as a result of structural movement acting, most usually from above or sometimes from below, when failure has occurred in the plinth as a result of either core compaction or differential ground settlement. Most structural cracks are more or less vertical but can run diagonally in cases where, for example, the plinth has failed, allowing a large section of material at the foot of the wall to become detached. Some cracking, which may at first sight appear quite serious, might actually be fairly superficial. For example, the block-like pattern of cracking which results from the use of clay-rich soils in rammed earth construction.

Although not an immediate problem, such cracks can represent areas of potential weakness if structural movement does occur in the future. The same is also true, though to a lesser extent, of the horizontal joints between lifts in wet-placed, piled cob and mudwall buildings, where outward movement at the wall head can cause the wall to tilt, and fracture along the line of a horizontal joint located at or near first-floor joist level. Most structural cracks in cob or mudwall buildings develop at corners or, in straight sections of wall, below main roof trusses or where cross-beams are bedded in the wall; in fact, any area where the earth wall is supporting a point load may be susceptible to this type of failure. Before going on to discuss structural crack stitch repairs it should be stressed that such work should be carried out only when all the causes of structural movement have been identified and effectively dealt with, and the building as a whole stabilised.

Less serious vertical cracks, those where there is no evidence of recent or continuing structural movement and the wall appears to be generally stable, may not need to be stitched but can be dealt with simply by cutting back and filling. This can be achieved by slightly enlarging the crack, then undercutting the cavity from both faces in order to form a hexagonal cross-section. This will provide a key for the repair, helping to prevent any future outward or differential movement (see Fig. 7.3g). The enlarged crack is filled from both sides at the same time, hammering or ramming into the pre-wetted cavity a damp earth/straw mix, as described above for the repair of hollows and cavities. Again, the new material should be left projecting about 15 to 20mm from the wall face.

The more serious cracks, in particular those located at the corners of buildings or which have been widened as a result of severe weathering and erosion, may be dealt with as follows. The first step is to carefully cut out horizontal chases (using a disc cutter and either a scutch or lightweight powered chipping hammer) of an appropriate size to accommodate the new repair material. Chases would be at vertical intervals of between 1.0 and 1.5 metres both externally and internally (see Figs 7.3 and 7.4). Chase dimensions will vary according to wall thickness, the materials to be used in the repair and their unit size, the thickness of mortar joints and the type of joint reinforcement, if any, it is planned to use. When repairing a cob or similar wet-placed mass earth wall, the thickness of which will be around 550 to 600mm, cutting out chases up to 200mm deep should not place the structure at risk.

However, in more slender walls, some of which may be no more than 400 to 450mm thick, a different approach may be required in order to avoid the removal of excessive amounts of material. For this reason an alternative method has been devised to deal with slender walls, which requires the cutting of chases no more than 100 to 125mm in depth. In this method the repair material, which may be either lime-stabilised earth or fired plain clay tiles bedded in lime mortar, is placed in vertical rather than horizontal layers or courses. Because this type of repair is inherently weaker than one using earth blocks or tiles laid in horizontal courses (described below) great reliance is placed on the incorporation of metallic reinforcement to provide the required tensile and flexural strength, and returns or rebates are located at both ends of the chase to further strengthen the repair (see Fig. 7.5).

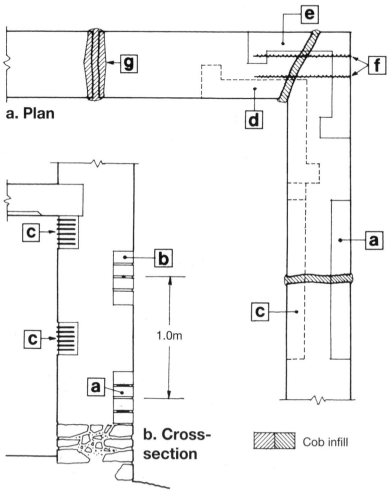

a. Plan

b. Cross-section

1.0m

Cob infill

Key to symbols
(a) Straight cob block stitch with eml reinforcement
(b) As above but with Helibar reinforcement
(c) Internal fired plain clay tile stitch with eml reinforcement
(d) Internal corner stitch with optional rebated ends
(e) External corner cob block stitch with rebated ends as above
(f) Inserted Helibar repair (for use in non-stony material)
(g) Filling of vertical crack with rammed cob

7.3
Types of structural crack stitches and their positioning

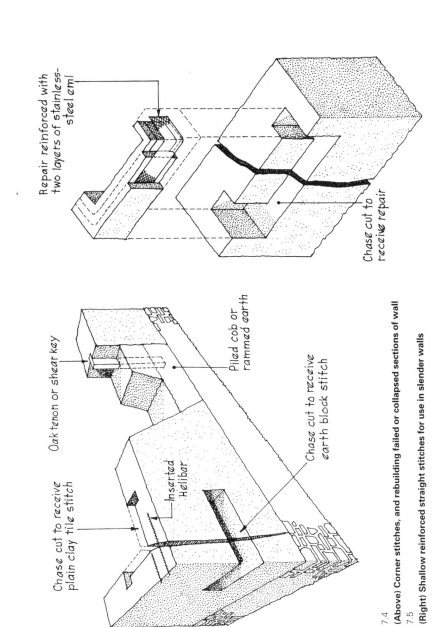

Repair reinforced with two layers of stainless-steel cml

Chase cut to receive repair

Oak tenon or shear key

Piled cob or rammed earth

Chase cut to receive plain clay tile stitch

Inserted Helibar

Chase cut to receive earth block stitch

7.4

(Above) Corner stitches, and rebuilding failed or collapsed sections of wall

7.5

(Right) Shallow reinforced straight stitches for use in slender walls

In Fig. 7.6 is shown a typical method of crack stitching using earth blocks. Earth/straw blocks were first used for structural repairs and for rebuilding sections of failed or collapsed cob walls some 15 years ago, and have been successfully employed for the repair of numerous historically important cob and mudwall buildings. Earth blocks and tiles, the fabrication of which has been described in Chapter 3, can be made to any convenient size to suit individual repair situations. For the thick (700 to 750mm) walls of some earlier cob buildings, concrete block size units (450 × 225 × 100mm) have been used for crack stitching but are generally considered to be more suitable for major rebuilding works. The width of blocks or tiles intended for use in crack stitching will be determined by the depth of the chase, which should be no more than one-third of the total wall thickness in order to avoid any risk of weakening the structure. In Fig. 7.6 the blocks illustrated measure 300 × 150 × 75mm. However, it may be considered that a unit size more akin to that of tiles or Roman bricks of, say, 350 × 175 × 50mm, might be more appropriate. If laid in four bonded courses, with stainless-steel expanded metal lathing (eml) reinforcement located below the top course and above the bottom one, the repair would require a chase no more than 300mm high

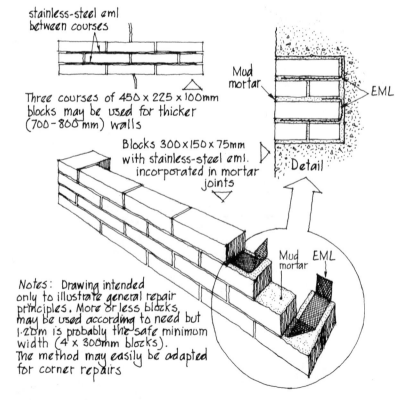

stainless-steel eml between courses

Three courses of 450 × 225 × 100mm blocks may be used for thicker (700–800mm) walls

Mud mortar

EML

Blocks 300×150×75mm with stainless-steel eml. incorporated in mortar joints

Detail

Mud mortar   EML

Notes: Drawing intended only to illustrate general repair principles. More or less blocks may be used according to need but 1·20m is probably the safe minimum width (4 × 300mm blocks). The method may easily be adapted for corner repairs

7.6
Some details of earth block crack stitches

to be cut out, which would mean having to remove less of the original material. The tiles would be laid in an earth mortar using a normal bricklaying technique, with staggered vertical joints (perpends) and with shards of slate wedged into the repair above the top course of tiles, thus providing a repair in which flexibility is combined with high tensile strength. Whether or not to include returns at either end of the repair, which it must be stressed should be no deeper than half the wall thickness, or to use metallic reinforcement in order to increase its strength, is a matter of choice, depending, among other things, on wall thickness, location of the repair and the extent to which the structure of the building as a whole has been effectively stabilised. For example, in straight sections of wall it may not be necessary to incorporate either returns or metallic reinforcement, but in corner repairs, where thrust on the wall, especially in the case of fully hipped roofs, will be acting in two directions, tending to force the corner apart, a stronger repair will be needed to counteract this tendency. A typical corner stitch repair, in which cob tiles have been used, is shown in Fig. 7.7.

For internal stitches, which are placed at similar vertical intervals to those placed externally but staggered so as to avoid weakening the structure, fired clay plain roofing tiles bedded in a lime/sand mortar may be used (see Fig. 7.3c and Fig. 7.8). In the repair method most usually employed, using tiles either 160 or 190mm wide, the following procedure is adopted. A chase 240mm high and 180 to 210mm deep, extending 70 to 90cm either side of the crack, is cut out, incorporating a return at each end of sufficient depth to accommodate the length of one tile. A minimum of six, more usually eight, courses of tiles are then laid in a mortar of 1:3 lime putty/coarse sand, with stainless-steel eml incorporated in every second or third joint. The eml reinforcement is folded over at the corners and bent around the back of the tiles in the returns at both ends of the repair, as shown in Fig. 7.6. Once the stitch repairs have thoroughly dried out, the remaining, exposed, parts of the crack may be filled using either an earth/straw mix, as described above, or, if the crack has been widened as a result of severe erosion, repaired using 300 × 150 × 75mm cob blocks bedded in an earth mortar. These blocks can easily be cut and trimmed to size and may be used in conjunction with a rammed cob mix to fill any irregularly shaped cavities.

The repair of severely eroded areas, rodent and plant damage

Severe localised erosion is normally found only in exposed, unrendered walls that have suffered long-term neglect: redundant farm buildings, for example, where erosion at the foot of an earth wall may have resulted from splash-back combined with the effects of driving rain as well as abrasion damage from animals and wheeled vehicles. Erosion higher up the wall is usually the result of driving rain, a leaking roof or lack of an effective rainwater disposal

7.7
**Repair of corner
crack and severe
erosion using cob
tiles**

7.8
**Fired clay tile
internal corner
crack stitch**

system. When severe erosion has occurred at the base of an earth wall, the masonry which serves to support it may also have been damaged or weakened as a result of the combined effects of rainwater and frost damage, leaching out the lime mortar and loosening or dislodging stones. The first step, therefore, would be to repair and stabilise this vital part of the structure so that the earth blocks, which will form the bulk of the repair, can be built off a firm and level base. The blocks are laid in a mortar comprised of fine earth screened through a garden or mason's sieve (5 to 6mm mesh). Mortar joints should be as shallow as possible in order to reduce the risk of settlement, normally 9 to 10mm, except where metal mesh or some other form of reinforcement is incorporated, when the joint might be 15 to 18mm in depth. The mortar should not be used in a very wet state as this will lead to shrinkage cracking in the joints. If the mortar contains a lot of clay, and is so sticky as to be unworkable, it can be gauged with well-graded sand. In order to avoid the removal of original material, blocks can be cut and trimmed to fit the cavity and keyed into the existing wall using Helifix wall ties. Spaces around and behind the blocks can be filled with a fairly stiff earth/straw mix well rammed into the remaining cavities (see Fig. 7.9).

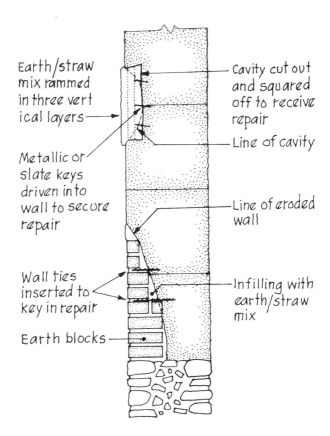

Earth/straw mix rammed in three vertical layers

Cavity cut out and squared off to receive repair

Line of cavity

Metallic or slate keys driven into wall to secure repair

Line of eroded wall

Wall ties inserted to key in repair

Infilling with earth/straw mix

Earth blocks

7.9

Repair of severe erosion at base of wall and repair of less serious cavities at the wall face

Rampant, uncontrolled ivy growth. Ivy growth of the type shown in Fig. 6.7, Chapter 6, can cause severe damage and disruption to both random stone and earth walls. Although the plant can be destroyed using a systemic weed-killer, this can take several months and the roots, which can penetrate the wall to a depth of 300mm, will, although dead, still remain intact. It is the removal of the roots that will cause maximum damage, leaving deep cavities in the wall that will need to be made good, in this, as in the previous case, by the insertion of an earth block repair. When preparing the wall for repair one must first remove all loose or damaged material and then form a level base upon which to bed the blocks (see Fig. 7.9). In order to key old material to new, the occasional block may be laid as a header and Helifix wall ties may be inserted between horizontal joints as work proceeds. Wall ties may be inserted into sound cob material above a course of blocks using either an ordinary hammer or a powered percussion drill fitted with an adaptor (available from the manu-facturer) at an angle of about 10 degrees above the horizontal. The exposed part is then bent down to form a horizontal reinforcement in the next mortar joint.

Rodent damage. Locating, with any degree of certainty, the depth and position of rat tunnels, or 'runs' as they are popularly known, and nesting chambers is quite a lengthy procedure, involving a certain amount of guess-work. Although it is possible to use sophisticated electronic equipment such as thermal imagers and infra-red thermometers (Pearson, 1992) the author has no experience of their use and cannot, therefore, comment on their effectiveness. The most commonly used method is to use a long power drill and, starting at an entry or exit hole, to trace the extent and depth of the tunnels. It has been sug-gested that the cavities that result from rodent damage can be filled by means of gravity or low pressure grouting. The problems associated with this method are, firstly, that one can never be entirely sure how effectively the cavities have been filled (they may, for example, contain all sorts of loose debris, including the desiccated corpses of rats) and, secondly, the introduction of large amounts of liquid into the wall may have the effect of further weakening it. The safest method to adopt, having first propped and shored the wall if necessary, is prob-ably to cut out sufficient material to expose the cavities, working in, say, one-metre lengths, and to ram in a stiff earth/straw mix, possibly stabilised with 6 to 10 per cent hydraulic lime, in vertical layers, allowing each one to become hard-ened off before adding the next. Alternatively, one could employ earth/straw blocks with a rammed-in earth/straw mortar used as infill material, ensuring that the top surface of the repair is securely wedged with slate or tile fragments.

The rebuilding of walls following major structural failure or partial collapse

When major rebuilding needs to be carried out, the method to be adopted will be dependent upon several factors, among which are: the location of the building and how accessible it is; the seriousness of the failure and which part of the building is affected, the time of year when the failure occurs; whether the building is statutorily listed or located within a designated conservation area; and, finally, whether the failure is the subject of an insurance claim, some form of litigation or perhaps a Dangerous Structure or Listed Building Repairs Notice, served by the local planning authority.

To illustrate the type of problems with which an architect, surveyor or structural engineer might be faced, one might describe a fairly typical case; hypothetical but based on experience of actual building failures. A two-storey Grade II listed cottage is located near the centre of a village, in a designated conservation area. The cottage was built at right angles to the principal road through the village, and its gable end wall immediately adjoins the pavement. In the middle of November, following a period of prolonged, heavy rain, serious fissures appear in the cement/sand render coat applied to the gable wall, which has also developed a noticeable bulge about 1.5 metres above pavement level.

Preliminary investigations suggest that the wall is very wet, that structural movement is ongoing and that there is a strong possibility of imminent collapse. In this situation some or all of the following actions will need to be taken. The wall will need to be shored externally and propped internally in order to make the building safe. This will require the pavement to be closed and cordoned off, half the road to be closed and, possibly, traffic lights to be set up, all of which would require the consent of various public bodies, including the local planning authority. An insurance claim would no doubt be submitted, and a structural engineer or building surveyor employed to carry out a detailed survey and prepare a repair specification. Clearly, in this situation everybody concerned is under considerable pressure to get the building repaired as soon as possible and, because there can be no question of waiting for the spring, when the weather will be suitable for mass earth building, the usual outcome is wholesale demolition followed by rebuilding with concrete or other cement-based blocks. Such repairs are, for the reasons explained earlier in this chapter, likely, in the long term, to adversely affect the structural integrity of the building; for this reason some alternative methods of carrying out urgent structural repairs using earth-based materials are briefly described below and illustrated in Figs 7.10A to C.

In cases where there is less sense of urgency, and the building concerned has good access as well as adjoining areas that can be used for the storage and mixing of materials, then conventional earth materials and methods may be employed. When preparing a collapsed or failed section of wall for rebuilding in mass earth, any material from the failed wall that is capable of re-use should be set aside and stored, under cover if possible. Remaining, intact sections of wall should cut back to form an even, sloping or vertical surface against which to place the new material. Stepped joints must be avoided, especially when a wet, piled method of construction is being employed, as this can result in the development of diagonal cracks as the material dries out. This would not, however, apply to rammed earth, where shrinkage and settlement would be less, 0.25 to 0.5 per cent as opposed to 1.5 to 2.0 per cent for well-compacted cob or mudwalling. In order to minimise the risk of subsequent movement at the interface between old and new material, in vertical or near-vertical joints an oak tenon, or shear key, may be installed (see Fig. 7.4). A vertical slot is cut into the exposed existing wall surface and the shear key is then placed in position, set in a lime/sand mortar and fixed to the existing wall by means of wall ties or oak pegs lightly driven into pre-drilled holes (never try to hammer an oak peg into dry cob). Any new or replacement openings can be built in as work proceeds, using the methods described in Chapter 4 for new building. In cases where the roof remains *in situ* and intact or has been propped back into position, there may be insufficient space for mass walling to be built up to wall plate or eaves level. The usual solution to this problem is, having built up to a certain level in mass earth, to construct the remaining part of the wall with earth blocks. A new section of wall plate can then be fixed into the upper face of the earth blocks to support the main truss and rafter feet. Vertical settlement can be kept to a minimum by using a mix that is high in sand and low in expansive clays, and by allowing each lift to harden off and consolidate before adding the next; however, bear in mind that each successive lift will increase the compressive stress acting on those at the base of the wall.

In Figs 7.10A to C are shown three methods that might allow urgent repairs to be carried out during the winter months or when access to a building is so restricted that conventional mass earth-building techniques cannot be employed. The sketches are intended only to illustrate general repair principles and should not be regarded as being in any way definitive. All three were designed for use in repairs in cob or mudwall buildings but only Method 1 has been extensively used for major rebuilding works. Method 2 is based upon a repair developed by English Heritage, which has been successfully used on at least one occasion, while Method 3, considered to be the least satisfactory in conservation terms, is untried and therefore unproven. In Methods 1 and 2, extensive use is made of pre-formed earth blocks, the dimensions of which will be determined by existing wall thickness. In early buildings, whose wall thickness is around 675mm (27 inches) standard concrete block-sized units (450 × 225 × 100mm) have been widely used. However, for later eighteenth- and nineteenth-century walls with an average thickness of 550 to 600mm (22 to 24 inches) blocks made using a mould with internal dimensions of 400 × 200 × 100mm would be more suitable. The aim is to produce a one-and-a-half block wall, so that the blockwork courses can be overlapped as well as bonded (see Fig. 7.10A). A rather different approach is adopted in Method 2 (Fig. 7.10B) where a diaphragm or cellular wall is shown. For this method thicker, 150mm units placed on the edge could be used, laid in bonded courses, with the whole repair strengthened by the use of wall ties spanning the cavities and stainless steel eml laid in the joints of blocks used as cross-bonding elements. (It should be borne in mind that lifting and handling blocks of this size, unless they were of low density, i.e. 1,100 to 1,200kg/m^3, could have health and safety implications.) Cavities could, if required, be filled with lightly tamped earth or some form of lightweight aggregate. (This is a method that might well lend itself to the construction of new buildings, in order to comply with Building Regulation requirements relating to thermal performance).

With regard to Method 3 (Fig. 7.10C) there are, at present, no known examples of major reconstruction using composite materials and as a repair method it must be regarded as far from ideal unless, that is, stabilised earth blocks could be employed. However, as an alternative to total rebuilding in concrete blockwork it should not perhaps be entirely ruled out. An inner skin of 150mm concrete or lightweight blocks, or possibly stabilised earth blocks, with vertical pillars incorporated to act as buttresses and supports for structural timbers, could be constructed fairly rapidly. The blocks, which would need to be toothed into the existing earth wall at either end of the repair, would be laid in a hydraulic lime/sand mortar to obtain a set in winter conditions, otherwise a lime putty/sand mortar would be more appropriate. Stainless-steel eml placed across, and projecting from, the blockwork every second or third course would act as a bond to the outer section of earth wall. One advantage of this method is that the building could be made habitable quite quickly and the outer section of wall constructed at leisure. If necessary, a timber framework clad with corrugated iron, PVC or plywood sheeting could be installed to keep out the weather until conditions become suitable for mass earth construction.

The outer, mass earth section of the repair could be constructed using either shuttered rammed earth, which would dry out faster and be less susceptible to frost damage, or well-consolidated cob made from stable (non-expansive) subsoil. In either case the blockwork inner skin would act as a form of permanent shuttering against which the earth could be rammed or compacted. The stainless-steel eml left projecting from the inner block wall face would be laid between the lifts of rammed earth or cob in order to bond the two sections of wall together. The use of a lime-based mortar, together with the flexibility of the eml cross-bonding, should ensure that any potential problems from differential settlement or thermal movement are minimised. It should be noted that a hydraulic lime/sand mortar should be used only in conditions where the air temperature exceeds 5°C.

A

Joints reinforced with eml if required

Mud or mud/lime mortar

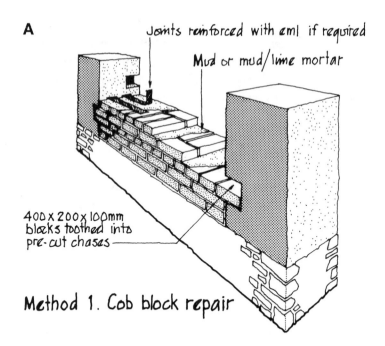

400 x 200 x 100mm blocks toothed into pre-cut chases

Method 1. Cob block repair

Existing cob cut back to form vertical faces (both methods)

B

Metallic non-ferrous wall-ties

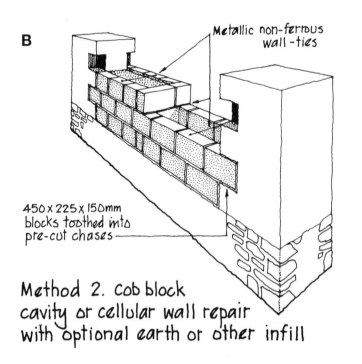

450 x 225 x 150mm blocks toothed into pre-cut chases

Method 2. Cob block cavity or cellular wall repair with optional earth or other infill

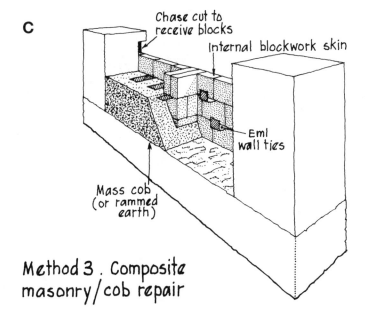

C

Chase cut to
receive blocks

Internal blockwork skin

Eml
wall ties

Mass cob
(or rammed
earth)

**Method 3 . Composite
masonry/cob repair**

In order to carry out repairs using Methods 1 and 2, large numbers of ready-to-use earth blocks would be required; therein lies a problem, because raw (unstabilised) earth blocks can normally only be mass-produced during the period April to September. It might be possible to produce blocks stabilised with hydraulic lime indoors during the winter months, but drying out and storage could still pose a major problem. Clearly, what is needed is a plentiful supply of suitable earth blocks, assuming someone could be persuaded to manufacture large numbers of them, fabricated to an agreed size and specification, either on a purely speculative basis or for pre-purchase and bulk storage. In Fig. 7.11 is shown a section of wall, parts of which had been severely damaged by ivy and where a failed stone facing had to be removed, entirely rebuilt in mass cob. Figure 7.12 shows a major structural repair carried out using cob blocks and Fig. 7.13 shows a collapsed cob wall being rebuilt using the same technique.

Repairing failed masonry plinths, underpinning, replacement of failed brick and stone facings, installation of land drains

When a stone or brick plinth has failed and the earth wall above remains (for the time being) intact, then it will need to be repaired in such a way that no further structural movement can take place. In cases where loose core material in a random stone wall has shifted and pushed the facing stones out, repair by pressure grouting, which may be considered as a possible option, would probably not be effective. Instead, repair by rebuilding would need to be carried out incrementally, dealing with one short length of wall (1.0 to

7.11
Failed wall rebuilt
in four lifts of
mass cob, also
showing cob
blocks used as a
facing to an
adjoining badly
eroded section of
wall

7.12
Structural repair using
450 × 225 × 100mm cob
blocks toothed into
existing wall

7.13

**Major rebuilding
of a failed cob wall
using cob blocks
bedded in an
earth/lime mortar**

1.5 metres) at a time. In order to avoid further disruption to the earth wall, which will probably already be supported by raking shores, it may be advisable to provide additional support from below using temporary 'needles' or dead shores, as shown in Fig. 7.2.

Facing stones are removed from the external face of the plinth, together with all loose rubble from the core of the wall, exposing the backs of some or all of the stones on the opposite face. The dismantled wall face can then be rebuilt, using new matching stones where necessary, and the core progressively filled with concrete. Stainless-steel wall ties are inserted at intervals to bond both faces of the wall together.

Underpinning, the possible dangers of which have previously been referred to on page 143, is something that should be considered only when all other options have been thoroughly explored. It is probably true to say that, in areas where earth buildings are commonly found, the sub-soils are likely to be fairly stable. However, even in these areas there will be situations in which ground movement may occur – the felling of mature trees and changes in water table levels for example, in which case either a geo-technical engineer or a structural engineer with experience of earth-building failures would need to be consulted.

Another problem sometimes encountered is where, at some time in the past, a badly eroded wall has been repaired by the addition of a stone or, more usually, brick facing. When these have failed or become detached from the earth wall, as they frequently do, then they will need to be taken down and replaced with an earth-based material securely keyed into the eroded wall surface, always provided that it has remained intact and in basically sound condition. If the wall has eroded back to a depth of less than one-third of its total thickness, it should be possible to re-face it using earth blocks, possibly containing lightweight aggregates, keyed into the eroded wall face, as described above for severely eroded areas.

Finally, one should mention the problems associated with raised ground levels and rising dampness described and illustrated in the previous chapter. When a building is situated at the foot of sloping ground on a clay-rich subsoil through which surface water cannot easily drain away, the obvious answer, provided that site conditions are suitable, is to install a land drain. Where the external ground level is higher than internal floor level it will need to be lowered to a new level at least 150mm below the internal floor, if necessary building a retaining wall to prevent any future problems developing. In Fig. 7.14 is shown a typical arrangement for improving drainage at the foot of

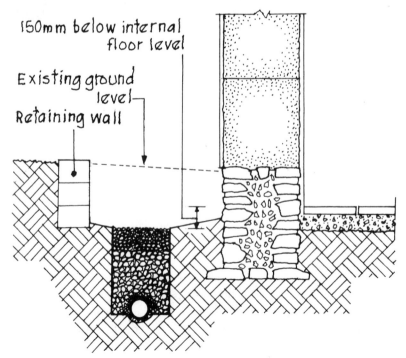

150mm below internal floor level

Existing ground level

Retaining wall

7.14
Alleviation of rising dampness by means of improved drainage

an earth wall. Trenches should ideally be located at least 300mm away from the wall in order to avoid disturbing the plinth and its foundation, if one exists. A 100 or 150mm diameter perforated pipe is laid in a depression at the bottom of the trench and surrounded by pea gravel. The trench is then back-filled with coarse gravel topped by a 150mm deep layer of fine gravel up to ground level. The pipes, laid to a suitable fall, are taken to a soak-away designed and sited in accordance with the relevant Building Regulations.

Organisations

British Geological Survey
Keyworth
Nottingham
NG12 5GG
Tel.: 0115 936 3488

Building Research Establishment
Bucknells Lane
Garston
Watford
Hertfordshire WD2 7JR
Tel.: 01923 664664

Centre for Alternative Technology
Machynlleth
Powys SY20 9AZ
Tel.: 01654 703912

Centre for Earthen Architecture
University of Plymouth
School of Architecture and Design
Hoe Centre
Notte Street
Plymouth
Devon
PL1 2AR
Tel.: 01752 233630

CRATerre-EAG
International Centre for Earth
Construction
Maison Levrat
Rue du Lac
BP53
F-38092 Villefontaine Cedex
France
Tel.: +33 4 74 95 43 91

Devon Earth Building Association
c/o Peter Child
South Coombe
Cheriton Fitzpaine
Crediton
Devon EX17 4PH
Tel.: 01363 866813

EARTHA
East Anglian Earth Buildings Group
Ivy Green
London Road
Wymondham
Norfolk NR18 9JD
Tel.: 01953 601701

English Heritage
Building Conservation and Research
23 Savile Row
London W1X 1AB
Tel.: 020 7973 3000

Historic Scotland
Longmore House
Salisbury Place
Edinburgh
EH9 1SH
Tel.: 0131 668 8600

ICOMOS UK
Earth Structures Committee
10 Barley Mow Passage
Chiswick
London W4 4PH
Tel.: 020 8994 6477

ITDG

Intermediate Technology
Development Group
Schumacher Centre
Bourton Hall
Bourton-on-Dunsmore
Warwickshire CV23 9QZ
Tel.: 01926 634400

National Soil Resources Institute

Cranfield University
Silsoe
Bedford
MK45 4DT
Tel.: 01525 863 266

SPAB

Society for the Protection of
Ancient Buildings
37 Spital Square
London E1 6DY
Tel.: 020 7377 1644

University of Bath

Department of Architecture and
Civil Engineering
Bath BA1 7AY
Tel.: 01225 826646

University of Kassel

Building Research Institute
Am Wasserturm 17
Kassel
D-34128
Germany
Tel.: +49 561 883050

Bibliography

Borer, P. and Harris, C. (1998) *The Whole House Book: Ecological Building Design and Materials*, Machynlleth, Wales: Centre for Alternative Technology.

Bouwens, D. (1988) 'Clay Lump in South Norfolk: Observations and Recollections', *Vernacular Architecture*, Vol. 19, 10–18.

BRE Digest 245 (1985) *Rising Damp in Walls: Diagnosis and Treatment*, Watford: Building Research Establishment.

BRE Digest 410 (1995) *Cementitious Renders for External Walls*, Watford: Building research Establishment.

Brunskill, R. W. (1962) 'The Clay Houses of Cumberland', *Transactions of the Ancient Monuments Society*, Vol. 10, 57–80.

Brunskill, R. W. (1987) *Illustrated Handbook of Vernacular Architecture*, London: Faber & Faber.

BS 648 (1964) *Schedule of Weights of Building Materials*, London: British Standards Institution.

BS 1377 (1990) *Soils for Civil Engineering Purposes, Part 2, Classification Tests*, London: British Standards Institution.

BS 8014 (1992) *Code of Practice for Assessing Exposure of Walls to Wind-driven Rain*, London: British Standards Institution.

Carter, M. and Bentley, S. P. (1991) *Correlations of Soil Properties*, Plymouth: Pentech Press.

Countryside Education Trust (1997) *House in Time* (brochure), Beaulieu, Hampshire: The Countryside Education Trust.

Dayre, M. and Kenmogne, E. (1993) 'Etude des transferts d'humidité, dans les blocs de terre crue compactée', *Terra 93 Conference Papers*, Lisbon, pp. 348–52.

DEBA (2002) *Appropriate Plasters, Renders and Finishes for Cob and Random Stone Walls in Devon*, Exeter: Devon Earth Building Association.

Dobson, S. (2000) 'Rammed Earth Building: present trends and future possibilities', *Terra 2000 Conference Postprints*, James & James (Science Publishers) Ltd, London, pp. 22–26.

DVL (Dachverband Lehm eV) (ed.) (1999) *Lehmbau-regeln; Begriffe, Baustoffe, Bauteile* [*German Earth-building Standards*], Weisbaden: Viewag-verlag.

Easton, D. (1996) *The Rammed Earth House*, Vermont: Chelsea Green Publishing Co.

Einteche, A. (1964) *Soil-cement: Its Use in Building*, New York: United Nations, Department of Economic and Social Affairs.

Glew, D. (2000) 'Mud-and-stud Construction: Compliance with Modern Planning and Building Regulations', *Terra 2000 Conference Papers*, London: James & James, pp. 312–18.

Goodhew, S. and Griffiths, R. (2005) 'Sustainable Earth Walls to Meet the Building Regulations', *Energy and Buildings*, Amsterdam: Elsevier BV Vol. 37, pp. 451–59.

Goodhew, S., Griffiths, R., Watson, L. and Short, D. (2000) 'Some Preliminary Studies of the Thermal Properties of Devon Cob Walls', *Terra 2000 Conference Papers*, London: James & James, pp. 129–43.

Greer, M. J. A. (1996) 'The Effect of Moisture Content and Composition on the Compressive Strength and Rigidity of Cob Made from the Soil of the Breccia Measures near Teignmouth, Devon', unpublished PhD thesis, University of Plymouth.

Harrison, J. R. (1984) 'The Mud Wall in England at the Close of the Vernacular Era', *Transactions of the Ancient Monuments Society*, Vol. 28, 154–74.

Harrison, J. R. (1999) *Earth – The Conservation and Repair of Bowhill, Exeter: Working with Cob*, English Heritage Research Transactions, Vol. 3, London: James & James.

Heathcote, K. A. (1995) 'Durability of Earthwall Buildings', *Construction and Building Materials*, Vol. 9, No. 3, 185–9.

Higginbottom, I. E. (1966) 'The Engineering Geology of Chalk', *Chalk in Earthworks and Foundations, Conference Proceedings*, London: Institution of Civil Engineers, pp. 1–13.

Hodges, H. (1989) *Artifacts: An Introduction to Early Materials and Technology*, London: Duckworth.

Holmes, S. and Wingate, M. (1997) *Building with Lime: A Practical Introduction*, London: Intermediate Technology Publications.

Houben, H. and Guillaud, H. (1994) *Earth Construction: A Comprehensive Guide*, Intermediate Technology Publications.

Hughes, P. (1986) *The Need for Old Buildings to Breathe*, Society for the Protection of Ancient Buildings, Information Sheet 4.

Hurd, J. and Gourley, B. (eds) (2000) *Terra Britannica: A Celebration of Earthen Structures in Great Britain and Ireland*, London: James & James.

Hutchinson, P. O. (1857) Letter to *Notes and Queries, London, Second Series*, No. 102, 12 December 1857, pp. 480–1.

Jacob, J. and Weiss, N. R. (1989) 'Laboratory Measurement of Water Vapour Transmission of Masonry Mortars and Paints', *Association for Preservation Technology Bulletin*, Vol. 21, Nos. 3/4, 62–9.

Jaggard, W. R. (1921) *Experimental Cottages: A Report on the Work of the Department at Amesbury, Wiltshire*, London: Department of Scientific and Industrial Research, HMSO.

Keable, J. (1996) *Rammed Earth Structures: A Code of Practice*, London: Intermediate Technology Publications.

Keefe, L. N. (1998) 'An Investigation into the Causes of Structural Failure in Traditional Cob Buildings', Unpublished MPhil thesis, University of Plymouth.

Ley, A. and Widgery, M. (1997) *Cob and the 1991 Building Regulations*, Exeter: Devon Earth Building Association.

Little, B. and Morton, M. (2001) *Building with Earth in Scotland*, Edinburgh: Scottish Executive Central Research Unit.

Machin, R. (1997) 'The Lost Cottages of England: An Essay on Impermanent Building in Post-medieval England', unpublished paper presented to the Vernacular Architecture Group, winter meeting 1997.

McCann, J. (1995) *Clay and Cob Buildings*, Princes Risborough: Shire Publications.

Middleton, G. F. (revised Schneider, L. M.) (1992) *Bulletin 5: Earth Wall Construction*, Sydney: CSIRO Division of Building Construction and Engineering.

Minke, G. (2000) *Earth Construction Handbook: The Building Material Earth in Modern Architecture*, Southampton: WIT Press.

Norton, J. (1997) *Building with Earth: A Handbook*, London: Intermediate Technology Publications.

NZS 4297, 4298 and 4299 (1998) *Engineering Design, Materials and Workmanship for Earth Buildings*, Wellington: Standards New Zealand.

Oliver, A. C. (revised Douglas, A. and Sterling, J. S.) (1997) *Dampness in Buildings*, Oxford: Blackwell Science.

Pearson, G. T. (1992) *Conservation of Clay and Chalk Buildings*, Shaftesbury: Donhead Publishing.

Rowell, D. L. (1994) *Soil Science: Methods and Applications*, London: Longman.

Salzman, L. F. (1952) *Building in England Down to 1540*, Oxford: Oxford University Press.

Saxton, R. H. (1995) 'The Performance of Cob as a Building Material', *The Structural Engineer*, Vol. 73, No. 7, 111–15.

Bibliography

Schofield, J. (1994) *Basic Limewash*, Society for the Protection of Ancient Buildings, Information Sheet 1.

Schofield, J. and Smallcombe, J. (2004) *Cob Buildings: A Practical Guide*, Crediton: Black Dog Press.

Shears, R. T. (1968) *Conservation of Devonshire Cottages*, Bideford: Gazette Printing Service.

Standards Australia (2002) *HB 195: The Australian Earth Building Handbook*, Sydney: Standards Australia International.

Talbott, J. (1995) *Simply Build Green*, Moray, Scotland: Findhorn Press.

Tibbets, J. (ed.) (1992) Correspondence in *El Adobero Newsletter*, South West Solar Adobe, Bosque, New Mexico, USA.

Trotman, P. (1995) 'Dampness in Cob Walls', *Out of Earth II Conference Papers*, University of Plymouth, pp. 118–25.

Trotman, P. and Boxall, J. (1991) *Laboratory Tests – Vapour Transmission of Limewashes*, Report No. W91/723, Watford: Building Research Establishment.

University of Bath (2002–4) Research project 'Developing Rammed Earth Walling for UK House Construction' (http://www.bath.ac.uk/rammedearth/review.pdf).

Vale, B. and R. (1993) 'The Untapped Potential of Low-Energy Building', *Town and Country Planning*, August 1993, 205–7.

Vale, B. and R. (2000) *The New Autonomous House*, London: Thames & Hudson.

Vargas Neumann, J. (1993) 'Earthquake Resistant Rammed Earth (Tapial) Buildings', *Terra 93 Conference Papers*, Lisbon, pp. 503–8.

Walker, B., McGregor, C. and Stark, G. 'Earth Buildings of Scotland and Ireland', *Out of Earth Conference Papers*, University of Plymouth, pp. 26–36.

Walker, P., Keable, R., Maniatidis, V. and Martin, J. (2005) *Rammed Earth: Design and Construction Guidelines*, Watford: BRE Publications.

Williams-Ellis, C. and Eastwick-Field, J. and E. (1947) *Building in Cob. Pisé and Stabilised Earth*, Country Life (reprinted by Donhead Publishing, Shaftesbury, 1999).

Wingate, M. (1985) *Small-scale Lime Burning*, London: Intermediate Technology Publications.

Index

adobe (mud brick) construction 7, 8, 62, 94–5
    bitumen stabilised 53, 64
    mechanized production 64
    traditional method 63–4
Amesbury, experimental buildings 24–5, 84,
    93
appropriate repair materials 165–7
Australia 45
    rammed earth construction 40, 66, 88, 111
    technical standards 122

Bath University, research 3, 122
Beaulieu, Hampshire 13
Board of Agriculture (reports) 126
bonding stones 132
'breathable' building construction 69, 144
Brunskill, R. W. 12
BS 1377 1990, Soil Classification Tests 32, 36,
    39, 66,156
Buckinghamshire 12, 23,37
building defects 131–51
    damage from external sources 146–51
    inherent 131–3, 160
    resulting from alterations 136–9, 160
    roofs and other building elements
        133–6,161
    walls 131–2, 161–2
Building Regulations 3, 30–1, 84, 103–21, 181,
    187
Building Regulation Approved Documents
    104–5
    A – Structure 107–9
    B – Fire safety 109–11
    C – Site preparation and resistance to
        moisture 69, 111–13
    E – Resistance to the passage of sound
        113–14
    L1 and L2 – Conservation of fuel and power
        114–21
    Regulation 7 – Materials and workmanship
        105–7

Building Research Establishment (BRE) 121
    BRE Digest 245, Rising Damp in Walls 156,
        162
    BRE Digest 410, Cementitious Renders 154
    Overseas Division 10
    Research 154
buttresses 108, 142–3

calcium carbide damp meters 154–5
Cambridgeshire 22
capillarity 152
capillary moisture movement 147
cavity wall construction, see also diaphragm
    walls 120
cement
    and lime compared 97, 153
    -based renders and plasters 97, 140, 144
    embodied energy of 4–5
    screeded floors 144
cement stabilisation of soils 52, 93
Centre for Alternative Technology (CAT) 4, 10,
    94, 119
chalk cob 27, 49, 93
chalk, rammed 24, 27–8, 49, 93, 129, 131
Chalk, Upper 35, 48, 81
    cohesion 49, 56
    compressive strength 49
    density 43, 49
    geology 35, 48
    porosity 48–9, 155–6
    regional distribution 48
chimney stacks and flues 81–4, 133–4
CINVA ram block press 67–8
clay and bool construction 20
clay characteristics 45–8
clay dabbins 20
clay lump construction 14, 22–3, 62–3, 120,
    131, 164–5
clay mineralogy 46–7
clunch 12,48
CO_2 emissions 5

Index

cob (mudwall) construction 12–13
  blocks 63–5, 95, 174–5, 178, 180, 183–4,
    185
  fixings and structural supports 76–82
  formation of openings 71, 73, 75, 78–9
  mixing 59–62
  shuttered 76–7
  'slow' and 'quick' processes 12–13, 18, 20
  tools and equipment 60–1, 71
  wall construction 71–81
combustibility, *see also* fire resistance 81, 110
compressed soil blocks 67–8, 94–5
concrete, as a repair material 78
concrete blocks 139
concrete lintels 78, 84
concrete strip foundations 67–8, 84
conservation of earth buildings 127–9
Conservation Area legislation 102, 128
control joints, in rammed earth walls 88
Cornish 'clob' building 26
critical moisture content 156–7, 168
Cumbria 20, 127

dampness in earth walls 143,151–7
  effects of 147
  measurement of 154–5
  penetrating 118, 144–5, 168
  rising 144
dampness survey 162–3
damp proof courses 69, 112, 143–4
damp proof membranes 112, 144
Dangerous Structures Notice 179
dead shores (needles) 185
Devon 13,14, 16, 24, 26, 27, 28, 60, 125–7
diagnostic survey procedure 158–64
diaphragm (cellular) wall construction 181
Dorset 24, 26, 27, 76, 116, 126
Driving Rain Index (BS 8014) 160

East Anglia 14, 22–3, 94, 96
Easton, D. 66
Eden Project. Cornwall 94
electrical conductance damp meters 154–5
electron microscopy 45–6
embodied energy 2, 4
energy efficiency 2
energy consumption 2, 4–5
English Heritage 188
equilibrium moisture content 38, 152, 153
erosion and abrasion damage 150–1
European Union Energy Performance
    Directive 121

European Union technical approvals 121
expanded metal lathing (eml) 84, 165, 174–5,
    181

finishes for earth walls
  chalk slurries 29
  coal tar/sand 29
  limewash 29, 97, 100
  potassium silicate paint 100
  proprietary emulsion paints 144, 154
fire resistance, *see also* combustibility 110, 115
  BS 476 – Combustibility of building
    materials 110
  BS 6336 – Fire tests 111
  DIN (German) Standard 4102, Part 1 110
fired brick
  facings to earth walls 23, 26, 118, 158
  as a repair material 138–40
fired plain clay tile, as a repair material 171–2,
    175–6
fixings in earth walls, *see* cob (mudwall)
foundations
  for new buildings 67–70
  movement in 132–3
freeze/thaw conditions 156
French *pisé* building 90

gamma ray spectrometry 153
Germany 9, 88, 93, 122
Grenoble University, research 153
ground movement/differential settlement 146
ground levels, raised 146, 186
grouting
  rat runs 178
  failed masonry plinths 183

Haddenham, Buckinghamshire 23–4, 25
Hampshire, chalk buildings 16, 24, 26, 27
Helifix wall ties 76, 166, 169, 177, 178
Housing (Rural Workers) Act, 1926 126
Howard, Alfred 60, 78
humidity (internal), regulation of 97

Intermediate Technology Development Group
    (ITDG) 4, 10
Ireland 16–18
ivy damage 146, 178

Jaggard, W.R. 24, 84

Kassel University (Germany) research 62
Keim Granital paint 100

land drains 186–7
Leicestershire 22
lightweight aggregates 97, 120
  mineral 119, 120
  organic 114–5, 120
lime 4, 5
  hydraulic 53, 166
  hydrated 53
  NHL 2 (European Standard) 95
  putty 53
lime stabilisation of soils 52–3, 134, 166
Lincolnshire 14, 21, 119, 125
lintels 78–9, 82
Listed buildings 128, 179
Listed building legislation 102
Listed Building Repairs Notice 179
load-bearing capacity of earth walls 109

Macaulay Land Research Institute 32
macro- and micro-pores, see also soil density
  152, 155
maintenance of earth buildings 145–6
masonry plinths, see also underpin course
  158, 159, 162, 167, 177
mass earth construction
  wet, piled method 53, 57
  compaction method 53, 57, 65, 90
materials testing 106
moisture
  movement 139, 144, 153
  excess 144–6, 159, 165
  measurement (wet and dry weight) 151
mortars
  earth 94–5,170, 177
  lime 95, 175, 181
  composite 97–9
mud and stud construction 14, 21, 119
mudwall construction, see cob

National Soil Resources Institute 32
natural building materials 2–3, 106
New Zealand 122
Norfolk 22, 120
Normandy, France 99
Northamptonshire 12, 22

oak 133, 138
  for new construction 76
  for repairs 165, 180
out-gassing, of building materials 2
outward leaning walls 135, 142,161,
  164

optimum moisture content, see also Proctor
  test 66, 88
organic fibre reinforcement 57–8
  animal hair 96–7, 99
  chopped hay 97, 99
  straw 60, 132
Oxfordshire 16, 23, 24

passive solar design 120
permeability of applied finishes 153
Permo-Triassic geological formation 26
phenolic foam insulation 119
pisé de terre 14, 90
planning legislation and policy guidance 101–3
plasters and renders
  application of 99–100
  bond strength (adhesion) 96–7, 154
  cement/sand 97, 144–5, 165
  composite, earth/lime 97–9
  earth-based 96, 170
  gypsum-based 140
  lightweight 115, 120
  lime/sand 96–7, 144
plasticity index, see soil consistency
porosity, see soil density
Portland beds, geological formation 12
Plymouth University, research 40, 41, 58, 115
Proctor (optimum moisture content) test 66
puddled clay 13, 76
pumice stone (aggregate) 119

racking, in hipped roofs 134, 161
rainwater disposal 136, 145, 161
rainwater penetration 145–6, 161
rammed earth construction 14, 16, 66, 85–94,
  122, 132, 180
  formwork 85–91
  rammers and tampers 90–3
  traditional method 88–91
  vertical panel technique 87–8
  vibrating plate compaction 93
recycling/recyclable materials 3, 5, 105
renders, casting or throwing of 99–100
renewable resources 5
repair, principles of 164–5
repair, appropriate materials for 139–42
repairs to earth walls
  hollows and cavities 169–70
  major rebuilding 178–183
  masonry plinths/underpin courses 183, 185
  minor surface erosion 170
  rodent and plant damage 178

severe erosion 175, 186
stabilising failed walls, propping and shoring
167–9, 179
structural crack stitching 171–2, 173, 174
rodent damage 132, 135, 146, 149–50
roof coverings
corrugated iron 137, 169
slate 136, 145
thatch 12, 18, 20, 22, 135–6, 145, 161, 169
tiles 136–7
roof trusses
'A' frame 9, 18, 134
cruck frame 8, 9, 20, 134
jointed cruck 8, 9, 133, 137, 138
triangulated 136, 169
roof types
gabled 134
fully hipped 80–1, 134, 136, 161
half-hipped 134

Scotland 20–1
shrinkage and settlement in earth walls 81,
132, 180
shear key, use of for repairs 180
site access and safety 70
slenderness ratio 108–9
smoke hoods, see Chimney stacks
Society for the Protection of Ancient Buildings
(SPAB) 143, 164, 170
Soil, aggregates
Sand 44
Silt 44–5
Stone and gravel 43
soil analysis and testing 34–43, 163–4
clay fraction (sedimentation) 32–3, 36–7
compressive strength 34, 40–3
consistency (Atterberg limit test) 34, 37–9
density and porosity 34, 43, 152
expansiveness (swelling and shrinkage) 34,
39–40
grading 35–6
particle size distribution (psd) 32–3, 35, 36,
54–5
permeability (diffusivity) 152
settlement velocity 37
soil classification 31–2
soil compaction 53–4, 67
soil contaminants 112
soil mixing and preparation, plant and
machinery 56–7
soil modification 53–6
soil particle density 43, 152

soil stabilisation 52–3
soil types and characteristics 32
soil survey maps 32
soils, storage on site 56
soluble salts/hygroscopicity 155–6, 162
sound insulation 113, 120
Stokes' law see soil, settlement velocity
straw 57, 132
straw bales 57, 60
structural cracks 132–3, 135, 139–40, 161,
170
Suffolk 22
subsidence, see ground movement
sustainability 1, 121–2
sustainable development 102–3

technical standards and codes of practice
121–2
tensile strength
of fibre reinforced earth 58
of plasters 96
thermal characteristics of earth walls
conductivity (k) 58, 114–5, 120
inertia 114
mass/capacity 118–20
transmittance (U-value) 115–121
thermal insulation 116–19
tie rods 142–3
timber cladding, to external walls 118–9
Town and Country Planning Act, 1990
101–2

underpin course (masonry plinth) 11, 23,
67–70, 108, 112
underpinning failed walls 143, 185
USA, research 44
US Department of Agriculture 33
U-value calculation methods 116–121

voids ratio 152

Wales 11, 18, 119
waste materials 3
water vapour transmission tests 154
wattle and daub 11
Wiltshire 24
Winchester, Hampshire 24, 27, 28
witchert 12, 23, 157

Yemen 95, 107

zero/low carbon emissions 2